How to Master Wargaming: Commander and Staff Guide to Improving Course of Action Analysis

How to Master Wargaming: Commander and Staff Guide to Improving Course of Action Analysis

Table of Contents

Introduction	1
Visualization	7
Seeing Ourselves, the Enemy, and the Terrain	8
Historical Vignettes	9
Tools to Increase Proficiency in Visualization and Improve Course of Action Analysis	17
Conducting a Commercial Wargame	19
Choosing the Right Wargame	24
Wargame Mediums	25
Thoughts on Course of Action Analysis: Action, Reaction, Counteraction, and Adjudication	29
Course of Action Analysis Methods	33
Conduct of the Wargame (Action-Reaction-Counteraction)	37
Wargame Rules and Methods	38
Adjudication	40
Correlation of Forces and Means	42
Analog Correlation of Forces and Means and Step Losses	44
Conducting Action-Reaction-Counteraction Cycles	47
Wargame Vignette	50
Thoughts on Training the Staff	59
Example of a Staff Training Plan	63
List of Sample Commercial Wargames	65
Appendix A. Technical Assistance Field Team Task Guide: The Military Decisionmaking Process. Collective Task Number 71-8-5111	73
Appendix B. References	85
Appendix C. Supplemental Resources	89

File One: Course of Action Analysis (Wargaming) Videos	89
File Two: Professional Reading	90
File Three: Supplemental Training Material	93

Center for Army Lessons Learned	
Director	COL Christopher J. Keller
CALL Lead Analyst	Mr. Richard Averna
CALL Contributing Analyst	Mr. Donald N. Matcheck
Special Contributors	Dr. James E Sterrett, Directorate of Simulation Education, CGSC
	Mr. Michael B. Dunn, Directorate of Simulations Education, CGSC
	Dr. Bruce Stanley and the School of Advanced Military Studies

The Secretary of the Army has determined that the publication of this periodical is necessary in the transaction of the public business as required by law of the Department.

Unless otherwise stated, whenever the masculine or feminine gender is used, both are intended.

Note: Any publications (other than CALL publications) referenced in this product, such as ARs, ADPs, ADRPs, ATPs, FMs, and TMs, must be obtained through your pinpoint distribution system.

Introduction

> *"The general who wins the battle makes many calculations in his temple before the battle is fought. The general who loses makes but few calculations beforehand."*
>
> Sun Tzu
> *The Art of War*

Vignette: National Training Center, Present Day

Six days into a 14-day rotation at the National Training Center, the battalion had just completed its second force-on-force: A deliberate attack, breach, and assault. It had been a rough day.

The battalion commander made his way back to the tactical operations center, with his mission and the day replaying in his head: Clear Objective Omaha to facilitate the passage of an Armor Battalion Task Force in order to complete the brigade combat team's seizure of Objective Nebraska.

When the commander entered the tactical operations center, the S-3 was already back. The staff was around the map and the executive officer (XO) was thinking aloud. "The course of action (COA) had seemed simple enough, and the wargame had been a breeze."

The commander took off his helmet and sat down on a steel folding chair, sighing heavily. "That was ugly. It seemed almost from the beginning we were chasing the day."

The S-3 began to explain, "Yes sir, at about one kilometer across Phase Line Sprint, Alpha Company's (A Co's) lead platoon took two volleys of enemy tank fire, disabling two tanks and destroying one Bradley. That was a surprise. I was not sure where it was coming from. Then, we could see it was from Support by Fire (SBF) 1. SBF1 was just out of our range, so I set the section of mortars with us and put some suppression on it so A Co could begin bounding; they had to fight for the position."

"It seemed to be going okay, then one of the mortar tracks was hit with anti-tank fire. It looked like the enemy was pulling off the rise and A Co picked up its movement. As A Co occupied SBF1, its combat power was down to six M1s and three M2s. Mortar Section 1, down one track, was still en route to its position. Once on SBF1, we were fighting in two directions. It took us forever to gain any effects needed to launch the breach and I'm not sure we ever did. The enemy was more determined than in our wargame, and on top of that, made us fight for SBF1. We did not see that during the wargame, but we tried to address it during the combined arms rehearsal (CAR)."

1

The commander sat up. "Ok, one of the problems is right there. We have to get away from wargaming events at the CAR."

The XO nodded in agreement before picking up the explanation. "I was on the radio with the S-3 and the mortars used half their basic load and were down one tube. They were black on high explosives. A Co's expenditure of ammunition was higher than we wargamed. It forced us to shift some of the artillery fire to suppress the far right enemy platoon. That delayed the buildup of smoke on the main objective by about 40 minutes, or maybe longer."

The fire support officer muttered, "I am sure they will tell us in the after action review (AAR)."

The commander sighed, "How did that even happen?"

The XO continued, "Our wargame did not include the movement, just the actions at the objective. I was trying to save time so we could get the order out quicker. We assumed A Co could move rapidly to SBF1. Speed was one of our criteria, and we did not visualize the enemy forward of the obstacle. As a result, the enemy engaged A Co earlier, which caused A Co to occupy SBF1 with less combat power and without mortar support. A Co also lacked the ammunition to sustain direct fires that could destroy enough of the enemy on Objective Omaha."

The commander walked over to the map, "There is a second issue we cannot just wave off: the enemy. We have to assume it will fight, is free thinking, determined, and will try to surprise us and throw us off. That is what we have to wargame. Visualize the what-ifs."

The S-3 responded, "Yes sir, and from that event on, our timing was off. Shifting some of our artillery from the breach to compensate affected the high explosives and smoke supporting the breach. The smoke was dissipating when the breach was still ongoing and the enemy was not suppressed. We were losing tanks, and at that time, we had Bravo Company (B Co) send its tank platoon to reinforce Delta Company (D Co), and we had to commit Mortar Section 2 to support the breach instead of the assault. Our coefficient of forces, with one tank company and 45 minutes of artillery suppression and obscuration, won this in the wargame, but the enemy had voted and not in our favor!"

The XO thought to himself, "Our wargame just consisted of filling in the events on the synchronization matrix, and a review of the decision support matrix. Everything we did was perfect. The enemy died where and when we wanted it to. We thought if we missed something, we could catch it at the CAR."

The 3-291 Combined Arms Battalion (CAB) cleared Objective Omaha, but it was at a heavy cost. At the AAR that night, the cause and effect traced directly back to insufficient wargaming. The commander stood up and stressed what he saw as shortfalls in the wargame, saying, "We need to improve on the following: better understanding by the staff of its warfighting function, visualizing the fight with realistic adjudication of a thinking enemy, planning for and anticipating outcomes, synchronizing events and not just filling in a matrix, and finally, the CAR cannot be where we wargame."

WARGAMING

> "Wargaming is a disciplined process, with rules and steps that attempt to visualize the flow of the operation, given the force's strengths and dispositions, threats, capabilities, and possible COAs, impact and requirements of civilians in the area of operations (AO), and other aspects of the situation."
>
> Field Manual (FM) 6-0
> *Commander and Staff Organization and Operations.* 05 MAY 2014.

It is not the purpose of this handbook to repeat the doctrine on COA analysis. This handbook instead focuses on three items: First, how to improve and develop the cognitive skill of visualizing, a key component to COA analysis (wargaming); second, improving the methods and conduct of action, reaction, and counteraction adjudication of COA analysis with off-the-shelf wargames; and third, thoughts on training the staff.

Why focus on COA analysis? It is identified across all combat training centers (CTCs) and other training events as being the one step in the military decisionmaking process (MDMP) where staffs have the most difficulty. Further, integrating off-the-shelf wargames into staff training can help develop better understanding of how to apply rules and judgments to realistically adjudicate outcomes in the action, reaction, and counteraction process.

COA analysis (wargaming) is the fourth step in the MDMP and is arguably one of the most critical because it takes the commander's plan from concept to detail and synchronizes the unit's combat power for an operation. Recent observations of multiple units executing decisive action training environment rotations and Mission Command Training Program (MCTP) warfighter exercises (WFXs) have shown that battalions through division staffs do not conduct the wargame effectively. Staffs are challenged in visualizing events in time and space. Also, a solid home-station training plan for the staff on the MDMP is lacking. This leads to issuing incomplete

3

plans to subordinate units, a lack of shared understanding across the Army warfighting functions, and poor synchronization of all maneuver and support elements in the operation.

The following are observations from maneuver CTCs, MCTP WFXs, and training:

- Staffs are untrained on the MDMP upon arrival at the CTC. Typically, staffs will only have conducted one to two iterations of MDMP training prior to their rotations. Conversely, majors leaving the Command and General Staff College (CGSC) will have conducted eight to fourteen iterations of the MDMP before graduation. This number of repetitions could be a realistic goal for units prior to a deployment.

- Critical thinking can only be acquired in objective analysis — in this case doctrinal-based analysis (wargaming) — and applying that analysis to a tactical problem. More training repetitions and sets in staff exercises, command post exercises, planning drills, etc., are helpful. However, staffs feel they do not have time for repetitive training. Therefore, the CTC rotation becomes their single learning environment. Units do not turn enough repetitions to build an experienced staff team.

- Logic, reasoning, and visualization are the key cognitive skills that staffs require to conduct COA analysis. It is essential to visualize the fight in time and space. Generally, friendly and enemy forces have the same physical limitations. By understanding that concept, staffs can logically deduce what forces will need to accomplish their mission, and the decisions commanders will have to make.

- Only limited members of the staff understand the process of COA analysis (XO, S-2, S-3, and fire support officer). Unfortunately, many members of the staff do not understand their role in COA analysis, which links back to the MDMP training.

- COAs are far from complete and therefore, difficult to analyze. The enemy and friendly COAs are not understood by most of the staff going into the analysis step. Typically, often only the S-2 and S-3 are involved in the creation of these products. They are not staff-synchronized products.

- The chief of staff, deputy chief of staff for strategic plans (G-5), and XO do not allocate enough time to conduct COA analysis. By doctrine, one third of the planning timeline should be dedicated to COA analysis (the same time allocation as mission analysis).[2] Staffs will typically over-allocate planning time to mission analysis at the expense of COA analysis.

4

- Staffs struggle with "gathering the tools" prior to COA analysis. The most valuable tools are commander decision-making products, decision support matrices, and decision support templates.

- Staffs lack the doctrinal foundation to execute wargaming. Only the XO or S-3 may know the doctrinal foundation of COA analysis (wargaming), because they are Military Education Level 4. Staffs do not have a planning standard operating procedure (SOP) to address planning requirements for each warfighting function (WfF)/staff representative. Commanders are not often involved in the planning timeline to ensure their staff executes wargaming. Brigade staffs are also torn between current operations, future operations, and the plans cell. Therefore, planning horizons become muddled, and staffs are bouncing around and leaving wargaming to a skeleton crew. Staff noncommissioned officers are not involved in wargaming, but they are valuable assets and should be trained and integrated into the process. Staff NCCs have a lot to offer in finding on-the-ground friction when analyzing a COA.

The problem is training and educating unit staffs to become high-performance teams. The consensus in the field is that field-grade officers graduating from CGSC understand the academic steps of the MDMP to include COA analysis (wargaming). However, the dilemma in building any team is integrating individual members so they understand where and how they fit in the team, and how they contribute to the common goal. For example, the individual M240 machine gunner understands how to load, clear, take care of, and engage targets with his M240. However, he does not know how this weapon is employed within the squad to support the overall task and purpose. The machine gunner must learn how his squad leader thinks and acts, how the fire teams move, and what the SOPs and battle drills are, so he will know when to lift or shift fires, etc. Only after learning the SOP and battle drills, and practicing them during multiple field problems, first without and then with live ammunition, does he become an integral member of the team.

It is the same for a staff. Like any unit, the staff must train and educate its members individually and collectively, "…helping the commander understand the situation, make decisions, and synchronize those decisions into a fully developed plan or order … During COA development and COA comparison, the staff provides recommendations to support the commander in selecting a COA."[1]

In addition to training, there is a component of repetition or practice, which helps to hone the staff into a high performance team. Observations have shown that due to time, COA analysis is often rushed or not done at all. Like the adage "practice makes perfect," a staff that routinely rehearses the MDMP and continuously changes the conditions under which they will have to plan and generate orders, will be more likely to meet the high operating tempo of large-scale combat operations. COA analysis is similar to any collective skill, and is perishable if not continually trained and rehearsed. Therefore, it is the purpose of this handbook to provide thoughts on how to develop individuals and staffs so they can better conduct COA analysis during the MDMP.

This handbook includes supplemental resources providing additional information to assist in educating and training the staff. These resources can be accessed on the Center for Army Lessons Learned public website here: https://call2.army.mil/toc.aspx?document=17879 More information about these resources can be found in Appendix C on page 89 of this handbook.

Endnotes

1. FM 6-0, *Commander and Staff Organization and Operations*, 05 MAY 2014. Page 9-2.

2. Ibid. Page 10-4.

CHAPTER 1

Visualization

> *"Therefore I say: If you know the enemy and know yourself, you need not fear the results of a hundred battles. When you are ignorant of the enemy but know yourself, your chances of winning or losing are equal. If ignorant of both your enemy and of yourself, you are certain in every battle to be in peril."*
>
> Sun Tzu
> *The Art of War*

The ability to visualize events or activities that occur sequentially or simultaneously in time and space is a critical skill for military leaders and their staffs. The art of true visualization is to understand with clarity how an enemy can affect these events or activities. An American version of the preceding quote from Sun Tzu might be: See yourself, see the enemy, see the terrain. Army Doctrine Publication (ADP) 6-0, *Mission Command: Command and Control of Army Forces,* 31 JUL 2019, discusses the ability to create shared understanding, and how creating shared understanding is a challenge for commanders and their staffs.[1] In reality, it is a larger challenge for staff, because the commander's vision is already clear in his own mind. Although this chapter provides some perspective on how to better visualize the common picture, it offers no guarantee the reader will instantly become a Sun Tzu, Clausewitz, Patton, or Rommel. Developing this ability takes a lot of time, practice, and personal development.

ADP 5-0, *The Operations Process*, 31 JUL 2019, describes the commander's visualization as "...the mental process of developing situational understanding, determining a desired end state, and envisioning an operational approach by which the force will achieve that end state."[2] How can an individual or staff develop this cognitive skill of visualizing so they can improve the collective ability to analyze or wargame courses of action (COAs) to support the commander's vision?

SEEING OURSELVES, THE ENEMY, AND THE TERRAIN

We often know more about the enemy than we do about ourselves.

Seeing ourselves. Seeing ourselves is derived from all the combined running estimates. These estimates are part of the tools gathered and assessments made, which includes the training level of the force and leader experience. Each individual staff officer brings his own estimate as it pertains to his specific warfighting function (WfF). These estimates provide a snapshot in time, but are also part of the assumptions made about what the unit's combat power will be at execution time, projecting out 12, 24, and 48 hours. This projected combat power and associated combined arms capability is used for seeing ourselves during COA development and analysis.

A way. A crawl, walk, run approach: Each individual staff officer must become the subject matter expert (SME) of his WfF. Individual self-study and reflection is required. A way to improve visualizing single WfFs is to conduct a tabletop exercise (TTX). For example, engineers conducting a mechanical breach of a complex wire and mine obstacle. The engineer by himself or with his team (add the engineer commander if available) gets a piece of butcher paper, and puts grids on it. He then adds the obstacle, but nothing else goes on the butcher paper, not even the enemy. He lists out all the friendly engineer assets available and arrays them for the breach. He then lists all the critical events, and conducts the breach unopposed, taking notes of the time it takes. Next, he adds an enemy and conducts the exercise, changing the conditions and thus gaining insights and taking notes of friction points. This leads to a higher level of visualization of his WfF. Next, he will add terrain, then he will attrit his force, and so on.

"In the absence of definite information, units must be guided by their mission and by the terrain."[3]

Seeing the terrain. Seeing the terrain is vital. As Sun Tzu pointed out, if you know nothing about the enemy but you understand yourself, you have a 50 percent chance of winning. Visualizing how our forces or the enemy's forces will use the terrain to gain an advantage or exploit a vulnerability, is vital. By doctrine, the engineer owns the modified combined obstacle overlay (MCOO), which is essential for seeing the terrain. However, in reality every staff officer must be able to see the terrain, because it is crucial to developing the cognitive skill of visualization, which is the purpose of the TTX below.

> **A way.** The best way to visualize the good, the bad, and the ugly of the terrain is by getting out on it, walking it, standing on it, and discussing it. However, this is not always possible due to competing requirements in garrison. A TTX helps the staff build their ability to visualize terrain. To gain a better grasp of terrain, historical examples are key. For instance, Gettysburg. Maps of the area are readily available. Get the map, enlarge it if possible, array friendly and enemy forces as they were historically. Knowing the actions of both forces, focus on the terrain. See the advantages and disadvantages. There is a propensity to focus on the history, but the objective is to see, study, and visualize the terrain, and to gain a deeper understanding of how the terrain can be best utilized.

Understanding the enemy is not just the S-2's responsibility.

Seeing the enemy. Seeing the enemy consists of a detailed intelligence preparation of the battlespace (IPB) (i.e., considering the weather, terrain, disposition of the enemy, his strength, and likely COAs). This is the foundation for developing a COA, and thus sets the groundwork for an effective COA analysis. The S-2 is the overall SME on the enemy, but each staff officer must be an SME on the enemy's WfFs, known as reverse WfFs. Having this understanding is the first step in visualizing how the enemy will fight and employ its combat power.

> **A way.** This tabletop exercise involves the entire staff, led by the S-2. Start with an enlarged portion of a map (a MCOO of the terrain is essential). The S-2 develops the enemy maneuver and each staff officer then develops the reverse WfF enemy COA. The next step is to list out the enemy mission, task, purpose, and key events. Once complete, the staff then fights the COA — no U.S. forces — just a flow of the enemy fight. Crucial to this exercise are the discussions that should take place such as "Why is he doing that?" and "What are the effects?" It will be hard not to interject a U.S. force, but keep that to a minimum. The object here is to see the enemy and visualize its actions and effects. The more the staff as a team conducts this TTX, the deeper its understanding of the enemy will grow.

HISTORICAL VIGNETTES

The following historical vignettes illustrate how seeing ourselves, the enemy, and the terrain are essential in visualizing actions that lead to success. The first vignette is about Civil War Union general, BG John Buford Jr. at Gettysburg, and the second American Revolutionary War general, BG Daniel Morgan at the Battle of Cowpens.

**Figure 1-1. BG John Buford at Gettysburg
(Source: Library of Congress)**

BG Buford's defense of McPherson's Ridge outside Gettysburg is a classic example of the importance of visualization. On June 29th, Buford marched with his first and second brigades to Fairfield, PA. BG Buford knew a considerable size of enemy infantry was in the area. However, the exact location, size, and intent of the enemy was unclear. "The inhabitants knew of my arrival and the position of the enemy's camp, yet not one of them gave me a particle of information, nor even mentioned the fact of the enemy's presence.⁴" BG Buford arrived at Gettysburg on June 30th with less than 3,000 men. As his scouts entered the town, they skirmished with enemy patrols, who quickly withdrew west. BG Buford quickly deployed pickets, under the command of COL Tom Devin, covering the countryside and the roads leading in and out of town. COL Devin considered the rebel patrols insignificant, telling Buford he could handle whatever came along. However, BG Buford recognized the rebel patrols as being from two Mississippi Regiments of A.P. Hills Corps, of Robert E. Lee's Army of Northern Virginia. This was no small scrap. BG Buford's reply to COL Devin was clear "No, you won't. They will attack you in the morning, and they will come booming — skirmishers three-deep. You will have to fight like the devil until support arrives."⁵

BG Buford's years of combat experience taught him to hold the high ground. MG John Reynolds, with I Corps, was still a day or more away. BG Buford quickly understood his small force was the only thing that stood between Lee's army and Gettysburg. If he withdrew, Lee would gain the high ground, be set for the fight, and possibly deal a strategic blow to the Union Army. The remainder of June 30th proved to be busy for BG Buford as he surveyed the grounds, visualizing his force and the fight to come. "About a half mile west of the of Gettysburg town square is a moderate elevation called Seminary Ridge, running north and south and named for the Lutheran Theological Seminary that stands on its crest. This ridge is covered throughout its length with open woods. The ground sloping downward toward the west rises again to form McPherson's Ridge about 500 yards away. To the north, both ridges intersect at Oak Hill. West of McPherson is Herr Ridge and then Willoughby Run Creek crossing Chambersburg Pike."[6]

BG Buford realized the creek would keep the enemy in column on the Pike, causing a delay in their deployment. The three successive ridges gave him the high ground needed to fight a delaying action, and therefore buying him the time needed for MG Reynolds to close. He deployed his cavalry brigades as mounted infantry. Every fourth man stood to the rear holding horses, which effectively reduced his force to 2,200 men. However, dismounted, his men took up covered firing positions from behind trees, bushes, and fence posts, further maximizing the advantage of their breech-loading Spencer carbines. BG Buford positioned his six cannon for maximum effect. LT John Calef, specifically, " ... worked his guns deliberately with great judgment and skill, and with wonderful effect on the enemy."[7] On July 1st, BG Buford's Cavalry was hotly engaged, but by 1430, when MG Reynolds finally arrived, BG Buford had held the high ground and shaped the Battle at Gettysburg.

Insight. BG Buford possessed the ability to see the situation and to understand its second-and third-order effects. This ability might have come naturally, but likely it was because he had the training and the experience. In the 1850s he fought in the Indian Wars and had been serving in the Civil War up to this point, which had developed his ability to visualize his force, the enemy, and the terrain. In this example, BG Buford knew little about the enemy, only that he thought they would be coming down Cashtown Pike Road. He did, however, understand the terrain, visualizing the difficulty the enemy would have trying to deploy and that his force, small as it was, had the advantage of the Spencer Repeating Rifles, with a rate of fire of 14 to 20 rounds per minute. Fighting as infantry, his forces could delay the enemy, and therefore hold the high ground.

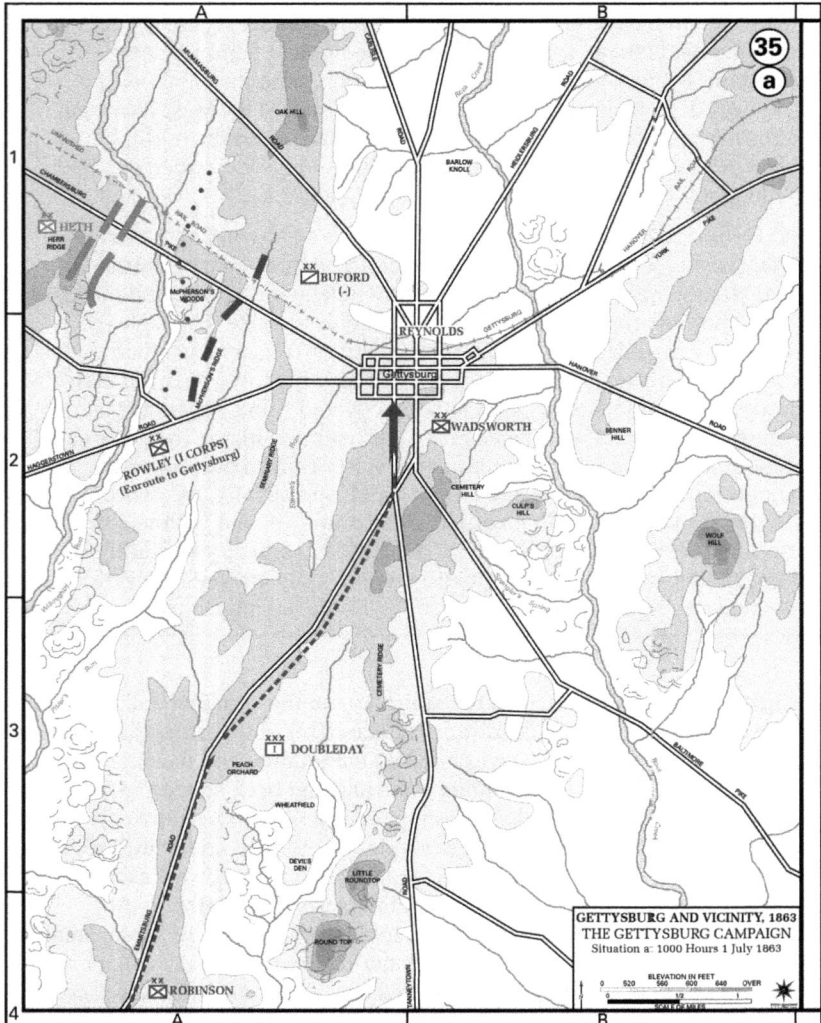

**Figure 1-2. BG Buford holding the high ground
Gettysburg, 1000 Hours, July 1863
(Source: United States Military Academy)**

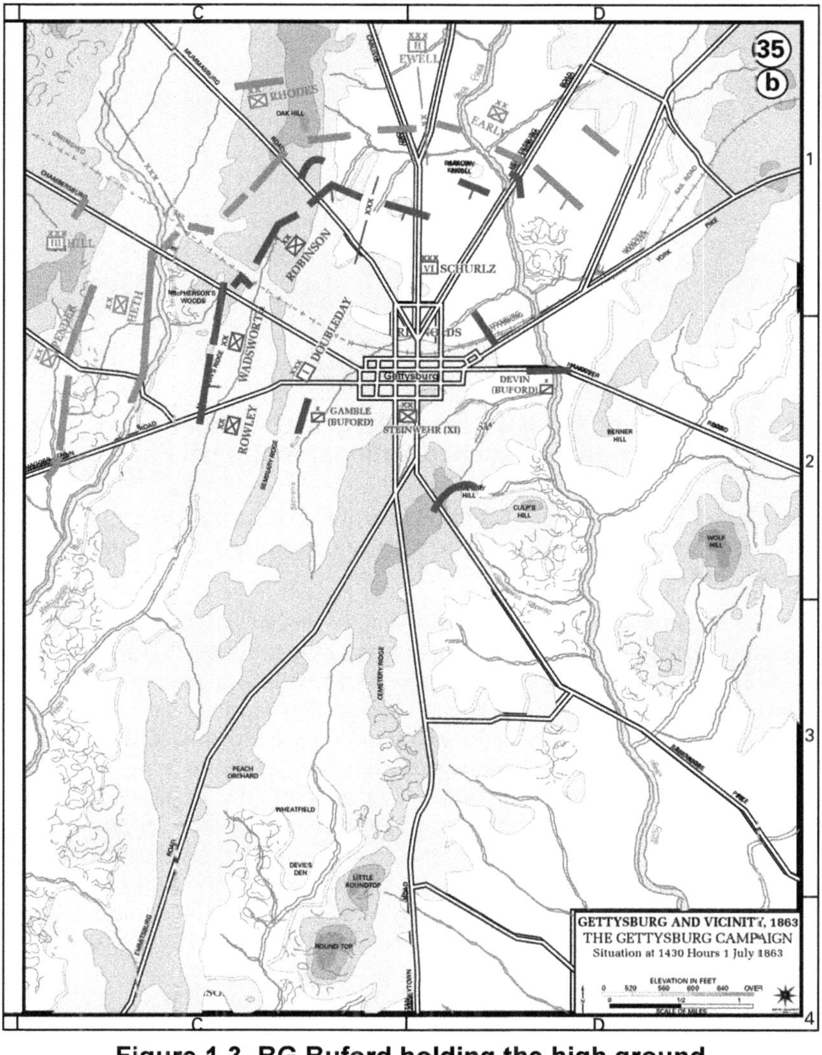

Figure 1-3. BG Buford holding the high ground Gettysburg, 1430 Hours, July 1863 (Source: United States Military Academy)

13

**Figure 1-4. BG Daniel Morgan at Cowpens
(Source: Independence National Historical Park)**

BG Daniel Morgan's actions at the Battle of Cowpens on January 17th, 1781 is an example of how to understand and use seeing the enemy, forces, and the terrain.

"When he led his party out of the wood at the end of Cowpens, Morgan halted to survey the ground in front of him to the northwest. The ground — meadow-like, as described by Major McDowell — sloped gradually upward to a low crest about 400 yards ahead. Beyond that was what appeared to be a ridge formed by two small hills. Morgan would later find that behind the nearer crest was a swale or extended dip running about 80 yards to the far or more northern crest. Taken in all, the terrain was indeed very gently rolling, with the higher ground never more than 25 yards higher than the plain. The rolling open terrain was ideal for the movement of cavalry, and there were no obstacles such as thick woods, swamps, or underbrush, which could serve to cover Morgan's flanks. In addition, the Broads River, about five miles distance, curved around the rear of the position, cutting off a retreat in that direction."[8]

BG Morgan understood his enemy. British Col Banastre Tarleton was a brash, young 27-year-old cavalryman who was out to make a name for himself. Morgan " ...was aware of Col Tarleton's impetuous charges and hell-for-leather tactics that could destroy an enemy caught by surprise."[9] Several of BG Morgan's officers had fought Col Tarleton before, and emphasized that he would not go in immediately, but hang back and send several troops in first. When Col Tarleton saw confusion or the enemy withdrawing, he would then charge with the reserve. Perhaps more importantly, BG Morgan understood the disdain that Col Tarleton had for the American fighting-man, especially the militia.

BG Morgan positioned his infantry in two lines on the open and sloping ground of the first hill. In the first line, he placed 150 militia in picket formation. They would lay low, engage the enemy at 50 yards, and then retire to the second line. The second line was 300 militia who were to fire two aimed volleys, then move off quickly to the rear and reform. The third and main line, his veteran Continentals, were positioned on the military crest of the second hill, hidden from the enemy. Half a mile behind the main line and concealed by the second hill, he positioned COL William Washington's Dragoons, reinforced by 45 mounted militia.

Seeing the American militia forward, and true to his reputation of not being able to restrain his eagerness, Col Tarleton ordered the attack before his commanders were ready. British infantry moved forward. The American first line engaged and fell back in order among the 300 militia. The militia firing two volleys momentarily stopped the British attack and then moved off to the left in "a river of men."[9] Col Tarleton saw a retreat and charged just as BG Morgan had envisioned. BG Morgan's main line blunted the charge and COL Washington's Dragoons caught Col Tarleton by surprise, driving him from the field. The veteran Continentals and the reorganized militia defeated Col Tarleton's committed reserve and the American force prevailed.

Insight. BG Morgan, unlike BG Buford, was not a career military officer. He did serve in the French and Indian War as a teamster, but it was during the Revolutionary War that he rose to become one of America's most competent generals. His actions at Cowpens make that point. His ability to see himself, the enemy, and the terrain was translated to victory. By controlling where he would fight, he negated Col Tarleton's ability to surprise him, offering the enemy limited maneuver space. BG Morgan's understanding of terrain allowed him to conceal his force in depth, turning a perceived weakness — his militia — into an advantage against Col Tarleton.

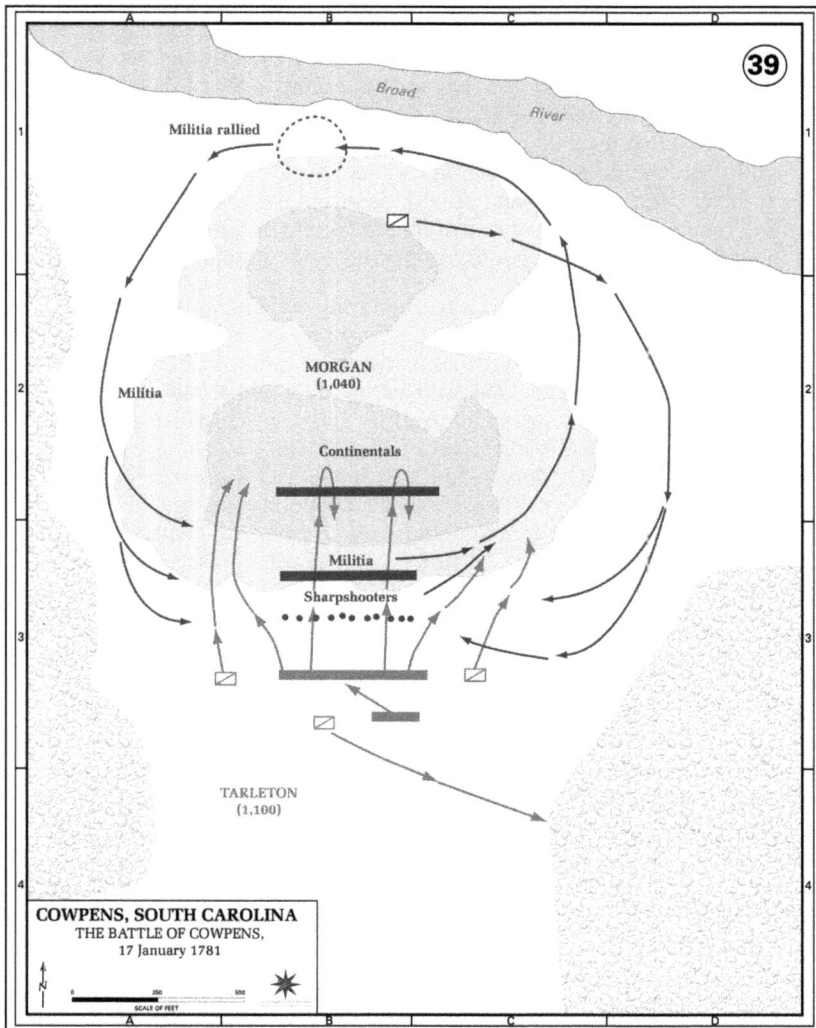

**Figure 1-5. The Battle of Cowpens
(Source: United States Military Academy)**

TOOLS TO INCREASE PROFICIENCY IN VISUALIZATION AND IMPROVE COURSE OF ACTION ANALYSIS

> *This is training for war! I must recommend it to the whole Army.*
>
> Field Marshal Karl Freiherr von Muffling,
> Chief of the Prussian General Staff

Commercial Wargames as Tools to Increase Proficiency

Commercial wargames are an excellent tool for professional development and increasing individual and collective proficiency. Wargames hone decision-making skills and visualization. When done in conjunction with a professional development program and home-station training plans, staffs can increase their proficiency as a team. **Note:** Commercial off-the-shelf wargames are specifically training aids, and should not be used to take the place of an actual operational COA analysis.

In the following two selections from a recent study conducted at the U.S. Army Command and General Staff College, *The Effects of Simple Role-Playing Games on the Wargaming Step of the Military Decisionmaking Process (MDMP): A Mixed Method Approach*, found in Volume 45 of *Developments in Business Simulations and Experiential Learning,* it was found that staff groups that played Kriegsspiel improved their visualization skills.[11]

Kriegsspiel

"The board game Kriegsspiel dates back to the early 1800s in Prussia, where it was used to teach members of the staff. Much of the game centers on a situation for each side, where they experience the 'fog of war.' Two sides play against each other, and are aided by running umpires that carry messages back and forth between the head umpire and player. Players see their unit represented on a map, with small blocks of different shapes and colors representing their role. During turns, the players write orders for their units, or updates for other friendly players. Every two minutes or so, the running umpire provides an update and collects new messages. The head umpire adjudicates movement and combat, provides outbound message traffic, and informs each participant's running umpire of what they can see and what their force is doing.

This manner of play forces the players, who are secluded from each other, to anticipate future requirements and analyze how to accomplish their mission alone. Explicitly, commanders learned that subordinate role players do not execute orders immediately because there is a delay for the travel of the message, and then a subsequent delay in the formation they command to respond to the order. Further, players gain an appreciation for concise mission-type orders, and an increased visualization of events in time and space.

Regarding arbitration, there are two forms. The original form, strict adjudication, is where combat is resolved by looking up force ratios, rolling a die, consulting a table, and accounting for each loss on a register. In the second type, accounting is simplified using ratios, tables, and accounting to keep the game moving at a brisk pace. The first provides for a more random outcome and the second provides a more likely outcome."

Developments in Business Simulations and Experiential Learning, p. 330

For Commanders and Staffs

"The chief recommendation, that resulted from the aforementioned study, was for commanders and staffs to wargame. Take the time to deliberately analyze COAs by wargaming as part of the military decisionmaking process (MDMP). All too often, planners skip the wargaming step entirely, and operations over the last ten years have only contributed to the atrophy of this skill in military planning. Wargaming provides commanders and staffs a method to analyze and compare COAs against one another, while testing the validity of the COAs against an uncooperative and thinking enemy. This test helps commanders and staffs identify gaps in planning, synchronize COA events in time and space, identify previously undiscovered threats and opportunities, and ultimately identify and think through potential decisions that commanders may be required to make in the execution of the fight. If planners skip or water down the wargaming step, then the gaps and synchronization will only become evident during rehearsals, or worse, in execution.

Wargame regardless of the number of COAs. When the commander gives a directed COA to staffs, there is only one option for execution. Therefore, in a time constrained environment, the perceived need to wargame only one COA may seem like a waste of time. After all, the sixth step of the MDMP is "course of action comparison," so why wargame if there is only one COA to compare? If COAs were perfectly developed, with no gaps in understanding or synchronization, then the wargaming step would seem to be a waste of time. However, no matter how skilled planners are, COAs can always be refined, and staffs must test them for the reasons identified in the paragraph above. Again, if staffs fail to wargame, then the gaps in understanding and synchronization will only become evident when it may be too late.

Consider role-playing games as staff and officer professional development activities in order to increase commander and staff visualization abilities. This study used Kriegsspiel, and found it to have a correlation with increased visualization, particularly with a planner's ability to better understand and visualize their own units on the battlefield. Other games and techniques may be useful to facilitate this end. Other options include GO™ (Ancient Chinese strategy game), Stratego™, Hunabi™, Chess, and simple visualization exercises of having subordinates draw out their understanding of the operation on a white board, or even in the dirt. This capability will help staff officers visualize operations, and enhance commander and unit understanding."

Developments in Business Simulations and Experiential Learning, p. 338

Wargames allow leaders to gain confidence in decision making through repetition and learning, improve visualization and understanding of military operations; and build teams and demonstrate individual personality traits and thought processes. Wargames allow experimentation in "safe-to-fail" environments and multiple opportunities to practice analysis and decision making. Wargames facilitate deep exploration of opportunity costs and incorporation of variables or conditions otherwise not easily replicable. Compared to most other training events, wargames require minimal overhead and resources. Many commercial wargames provide ready-made scenarios that are detailed enough to support basic planning drills. Most can be customized to create specific scenarios. If incorporated into a unit's history program, wargaming a historical event provides much greater depth of understanding than just reading about the event.

Time spent exercising the intellect to improve the understanding of warfare is time well spent. Wargames provide an essential ingredient to the well-rounded professional's education and training.

CONDUCTING A COMMERCIAL WARGAME

Like any training event, conducting a wargame requires time and dedication. To achieve maximum benefits, approach wargaming in a deliberate and disciplined manner. The following outline may seem intensive, but it provides planning, preparation, and execution, ranging from larger unit exercises like a command post exercise, to the list of off-the-shelf wargames listed at the end of this chapter. Most of these off-the-shelf wargames take minimal time to learn, set up, and play.

The primary components of a wargame event are learning objectives, targeted training audience (staff), setting and scenario, simulation, rules and adjudication procedures, supporting players and facilitation personnel, and assessment method.

Planning

1. Determine training objectives. What outcomes are desired? Examples include:

- Improving visualization in time and space

- Cohesion and team building

- Decision making in ambiguous circumstances

- Understanding and visualization of how units operate in time, space, distance, and uncertainty against an adaptive threat

- Understanding and visualization of how key leaders process information, make decisions, and execute the fight

- Understanding the technical aspects of the MDMP

- Understanding the standard operating procedures (SOPs) for how a unit conducts operations or the MDMP

- Developing expertise within individual duty positions, WfFs, and reverse WfFs, and their role in collective operations

- Understanding an operational environment

- Developing references and tools for staff estimates

- Familiarizing with a geographical area or a specific threat

- Practicing or exploring a specific event

2. Identify key events (such as the MDMP, reception, staging, onward movement, and integration [RSOI], delay, perform a wet gap crossing, etc.) to incorporate into the wargame:

- Identify the wargame to be used.

- Learn the wargame mechanics and components. The wargame is the vehicle to get to the training objectives. Understand both its opportunities and limitations. Identify any required modifications to the wargame, such as developing a customized scenario or rule modifications/workarounds. Ensure the selected wargame will do what is intended for the training.

3. Identify resources required:

- Personnel and responsibilities: Every player may have multiple exercise roles and duties. This is normal for small wargames. The following list is not meant to overcomplicate planning, but rather allow the planners to visualize wargame components better:

  o Exercise director. The exercise direction is in command of the exercise. He determines what is to be accomplished, approves how it is to be accomplished, and approves resources to commit, including time and participants. He also approves adjustments or time changes to key events, as well as additions, deletions, injects, or scenario adjustments.

  o Exercise planner. The exercise planner identifies and coordinates all required resources, and understands game mechanics. He develops scenarios and execution methods that meet training objectives, and assigns personnel their responsibilities for the exercise.

  o Simulation expert/facilitator. The simulation expert/facilitator understands all technical aspects of the game and conducts necessary training for players. He facilitates execution, and assists in adjudication. He also works with planners and exercise control (EXCON) to adjust scenarios and rules as necessary.

  o Exercise control. EXCON is responsible for conducting the exercise. They provide administrative instruction on scenarios, roles, responsibilities, and exercise conduct, as well as providing SME support for game play, ensuring game rules are understood, and adjudicating outcomes and any issues that arise. They also control the timeline, injects, and key events, and adjust as required.

  o Higher command. Higher command replicates the higher headquarters for players to interact with and receive orders/ guidance from. They often control adjacent units, or other units, to create the appropriate environment for the players.

  o Role players. These are non-competitive participants who interact with the training audience to create the desired environment, or stimulate desired behaviors.

o Observer/controllers. Observer/controllers observe and monitor execution and decision-making, records events for analysis and after action reviews (AARs), ensure EXCON is aware of events, decisions, and timelines, and identify training opportunities or frictions to EXCON. (They can also act as facilitators.)

o Support personnel. Support personnel may include tech personnel, security personnel, maintenance personnel, etc.

• White cell. The white cell resources, creates, and manages the environment and inputs for players to interact with the simulation, make decisions, and achieve the game's desired training outcomes.

• Blue cell. The blue cell, commonly known as friendly forces or blue forces (BLUFOR), are the active participants and training audience that function competitively.

• Red cell. The red cell is commonly known as opposing forces (OPFOR), and may or may not be actively competitive. They execute the scenario per guidance from EXCON. However, there is no reason why both blue and red cells cannot both be training audiences of equal stature and actively compete against each other.

• Gaming materials. How many copies of the game are needed? Any special requirements like rule summaries/cheat sheets?

• Computer game considerations. Computer games cannot be easily loaded onto government computers without proper authorization, and getting that authorization can be difficult and time consuming. What is the workaround? Use personal computers? Is it possible to network personal computers that are part of the design?

• Facilities. Are they secure and devoid of distractions? Do they support planning and briefings? Do they allow for private conversations out of OPFOR's earshot?

• Time. Is enough time allotted for set up, teaching rules, practice runs, scenario briefings and orientation, planning, execution, and the AAR/ hot wash?

Preparation

• Acquire gaming materials. Sometimes it is necessary to have multiple copies of the game.

• The game SME/exercise planner must learn game rules and conduct a game dry run to ensure the game is suitable to achieve training objectives. This person must know the game cold.

- The exercise planner and exercise director select the scenario(s).

- The simulation expert and exercise planner develop the additional resources that are required, such as scenario extracts, fragmentary orders (FRAGORDs), or gaming aids.

- Establish key events and identify any exercise control decision points, additional injects, branches, or sequels.

- Identify required game mechanic workarounds or facilitator adjudications. Some game rules are not worth the time or additional complexity to incorporate or adjudicate. Determine what is necessary and how to manually adjudicate without loss of training value.

- Game SME/exercise planner develops an appropriate method to teach participants enough rules to execute the wargame efficiently. These people do not need to fully understand every rule.

- Establish a timeline, which should include facility setup, teaching and practicing the game, scenario orientation, planning procedures, execution, and the AAR/hot wash. Some support personnel may require additional training and rehearsals.

- Set up facilities.

Execution

- A successful game is challenging, immersive, engaging, adversarial, and perceived as relevant. Make it fun!

- A game does not have to go to completion or have a clear front runner in order for learning to take place. Often, a better discussion will result if neither side has won.

- The key focuses for a successful wargame are, "...the players, the decisions they make, the narrative they create, their shared experiences, and the lessons they take away."[12] Also, focus on how the wargame shapes future understanding of warfare and decision making.

- Teach personnel how to play the game. Conduct a practice game in order to help them understand game mechanics.

- Conduct a scenario briefing and orientation.

- Conduct mission planning, even if it is a single person computer game. It is highly beneficial to go through the planning and decision process and brief plans.

- Facilitator/EXCON:

  o Keeps the game moving

  o Facilitates game play and rule execution

  o Adjudicates as necessary

  o Provides injects or guidance to players to meet training objectives, per guidance from the exercise director

  o Controls the exercise timeline

  o Records key events, decisions, etc.

Post-Exercise

- Conduct AARs.

- Update SOPs.

CHOOSING THE RIGHT WAR GAME

Wargames are tools. Like any tool, it is important to use the right one for the job. The first question should always be "What are the training and education objectives I am trying to achieve?" The next question should be "Does the game do what is necessary to address these objectives?" Any game design is about compromise. Some aspects will be abstracted, such as supply distribution or combat outcomes, in order to focus on some other aspect of the problem. Is the game relevant to the training exercise? Does the game do what you care about? The list at the end of this chapter provides some possible commercial wargames that can be used to train the staff.

Factors to consider:

- Is the game so complex that it will be difficult to learn and will bog down play?

- Do existing scenarios meet the objectives? How difficult is it to customize scenarios for the needs of the staff?

- How long does it take to play?

- What is the planning effort required?

- How many support personnel are required?

- What are the logistic considerations?

- What are the costs?

Often, games are rejected for suitability because they are not geared toward the "correct" echelon, but relevant learning will occur no matter what the echelon of the staff. Likewise, fixating on a "current environment" game unnecessarily discards many other excellent teaching games. Historical games can provide relevant experience, placing the participants in the shoes of those who wrestled with complex, ambiguous problems, and deliver a fantastic learning experience. Moreover, decision making with imperfect information under pressure does not fundamentally change era to era.

Keep in mind that many commercial games that are not military in nature can be surprisingly useful for team building, IPB, critical thinking, analysis, and decision-making.

Commercial gaming sites often provide in-depth reviews, tutorials, forums, and other resources to help make the best choices and assist with planning and execution.

WARGAME MEDIUMS
Computer Games

Advantages:

- May come with off-the-shelf, pre-loaded scenarios, requiring little setup time

- Automated bookkeeping, combat outcomes, etc.

- May facilitate fog of war

- Can be easy to learn

Disadvantages:

- The Department of Defense (DOD) generally does not allow commercial gaming software to be installed on DOD computers without an involved process. Privately owned computers often must be used. There might also be issues with networking on DOD systems.

- Multi-player participation may require multiple copies of the same game.

- May have limited scenarios or ability to customize. Adjudication is often in a black box, and thus participants may not understand why outcomes occurred.

- Limited ability to change or adjudicate outcomes or environments during execution to facilitate training objectives or explore branches and sequels.

25

• Limited ability to remove or mitigate onerous, unwanted, or distracting aspects of the game.

• May not allow multi-player teams or competitive head-to-head interactions; OPFOR may be controlled via artificial intelligence.

• Learning menus and interfaces may not be intuitive or easy.

• Fixation on computer terminals may inhibit learning discussions between players. However, the use of a proxy projector can allow for the image to be projected so a larger training audience can see it.

Board Games

Advantages:

• Encourage socialization, discussion, teamwork, and competition.

• Scenarios and rules can be truncated or adjusted to meet training objectives before or during execution. Branches, sequels, and "what-if" events can be explored.

• A single simulation expert/facilitator with mastery of the rules is often enough to run the game; all others need only be familiar with some key concepts.

• Can include fog of war with additional resourcing.

• System is open and participants understand why outcomes occurred.

Disadvantages:

• Setup can be tedious if there are many parts and pieces.

• Requires a simulation expert.

• Manual bookkeeping, combat outcomes, etc.

• Some boards and game pieces are not aesthetically pleasing and may seem unanimated.

• Fog of war may be resource intensive to incorporate.

• May be difficult to store if played over multiple days.

Miniature Games
(Scaled three-dimensional terrain models using micro armor [i.e., Dunn-Kempf])

Advantages:

- Visually engaging and immersive. Games with realistic units and terrain generate almost irresistible interest.

- Allows better visualization of the terrain.

- Rules and scenarios can be adjusted to meet training objectives before or during execution.

- Encourages contemplatively standing around the map board, even during pauses.

- In most cases, a single facilitator is adequate to have mastery of the rules; all others need only some key information.

Disadvantages:

- Terrain and model pieces may be quite expensive and time consuming to assemble.

- Replicated area may be limited.

- Rules can be complex.

- Requires a large area to set up.

- May be difficult to secure/store if played over multiple days.

- Fog of war requires some creativity.

WARGAME MISTAKES

- Failure to differentiate wargames from reality: Wargames are not completely realistic simulations. They model specific aspects of reality, but in doing so invariably give up other aspects of realism. The precise technical replication of real-world outcomes may or may not be accurate. For example, artillery may or may not be as effective as the game projects.

- Failure to use preparation for the game as a formal training opportunity unto itself: Do not jump into a game, use it to practice analysis and planning.

- Failure to understand the rules and conduct of the game: It is good to spend time on technical rehearsals, so the rules and mechanics will not become a distraction during execution.

- Failure to make realistic decisions ("gaming the game"): Avoid doing activities that would not happen in reality, or that real troops would refuse to do.

- Obsessing over game mechanics and rules: Make a plausible decision or outcome and get on with it.

- Failure to dedicate enough time for the event

- Failure to shield the event from distractions

- Failure to conduct an introspective analysis and AAR after the game

Note: For a list of sample commercial wargames, see Chapter 3.

Endnotes

1. ADP 6-0, *Mission Command: Command and Control of Army Forces*, 31 JUL 2019.

2. ADP 5-0, *The Operations Process*, 31 JUL 2019. Pages 1-8

3. Infantry School Staff. (1939) Terrain. *Infantry in Battle* (pp. 69-78). Washington D.C.: The Infantry Journal Incorporated.

4. LTC Smith, C. R. (1863). Brigadier General John Buford, U.S. Army, commanding First Division: Battle of Gettysburg. In United States War Department (Ed.), *The War of the Rebellion: a compilation of the official records of the Union and Confederate armies.* (Series 1, V. 27–Part 1) (pp. 927-993). Washington, Government Printing Office.

5. Longacre, E.G. (2000). *Lincoln's Calvarymen: A History of the Mounted Forces of the Army of the Potomac.* (p. 183) Mechanicsburg, PA: Stackpole Books.

6. Stackpole, E. J. (1982). *They Met At Gettysburg.* Harrisburg, PA: Stackpole Books.

7. LTC Smith. Ibid.

8. Wood, W. J. (1990). *Battles of the Revolutionary War 1775-1781.* Chapel Hill, NC: Algonquin Books of Chapel Hill.

9. Wood. Ibid.

10. Wood. Ibid.

11. McConnell, R. A., Gerges, M., Dalbey, J., Dial, T., Hodge, G., Leners, M.,... Schoof, P. (2018). *The Effect of Simple Role-Playing Games on the Wargaming Step of the Military Decisionmaking Process (MDMP) A Mixed Method Approach.* Fort Leavenworth, Kansas: U.S. Army Command and General Staff College.

12. Ministry of Defence. (2017). *The Wargaming Handbook.* The Development, Concepts and Doctrine Centre: Ministry of Defence Shrivenham.

CHAPTER 2

Thoughts on Course of Action Analysis Process: Action, Reaction, Counteraction, and Adjudication

> *"Victorious warriors win first and then go to war, while defeated warriors go to war first and then seek to win."*
>
> Sun Tzu
> *The Art of War*

This chapter provides thoughts and ideas on course of action (COA) analysis, focused on action, reaction, counteraction, and the adjudication process. COA analysis is the step in the military decisionmaking process (MDMP) that links COA development to COA comparison and COA approval. COA analysis facilitates visualization and understanding of the fight in time and space, allowing commanders and staffs to determine the optimal COA, and to identify difficulties, coordination problems, and probable consequences of the planned actions for each COA being considered. It also facilitates detailed planning and allows synchronization of warfighting functions (WfFs). Within the MDMP, COA analysis is often referred to as wargaming.

COA analysis is used to visualize the battlefield through the operational framework of deep, close, and consolidation areas over time and space. Echelons above brigade focus on intelligence, fires, protection, and sustainment to set conditions for the maneuver close fight, and to manage transitions to the next phase. Brigade and echelons below focus on visualizing and synchronizing the close fight.

A successful COA analysis:

• Is structured and rules-based

• Is a combination of science and art

• Is honest, introspective, and objective

• Applies critical thinking and avoids bias or mind traps

• Explores second-and third-order effects

- Includes realistic and plausible:

  o Environments

  o Friendly and enemy forces and capabilities

  o Decisions

  o Adjudication of actions

- Has detailed notes recorded

- Updates the products/estimates throughout

The following are tips for a successful COA analysis:

- Have all tools on hand, prepared, and ready.

- Have all the correct participants in the correct places.

- Have an ergonomic set-up. Participants must be able to see, hear, and be free from distractions.

- The facilitator must clearly understand and be able to articulate the time period and what events will be covered in each action-reaction-counteraction segment.

- All participants must understand exactly what is to be wargamed in time and space.

- Understand the common thinking or psychological traps, biases, and prejudices that can create inaccurate assessments. Examples include being emotionally attached to a COA, confirmation bias, or anchoring. Even a participant's branch, rank, or personal experience can create unconscious bias.

- Be honest and candid when evaluating a COA. Commanders must create an environment where subordinates can freely and professionally voice concerns and disagreements, especially with those who outrank them or have domineering personalities.

- Ensure adjudication outcomes are realistic and probable. Focus on the most probable outcomes; less probable outcomes should be wargamed separately as branches.

- Focus on decision points and explore second-and third-order effects and mitigations.

- Each participant should understand their own inputs and outputs to the wargame, and how to succinctly present them.

- Participants should provide input into reverse WfFs and specialties to assist intelligence in describing environment and enemy actions/capabilities.

- Identify advantages, disadvantages, risks, and areas requiring further study as the wargame progresses.

- As the wargame progresses, fill out, develop, and refine additional products such as tasks to subordinates, coordinating instructions, and additional graphic control measures, decision support templates (DSTs), and synchronization matrices.

- Include specialty liaison officers (LNOs) in the planning process. These might include attachments that are unfamiliar with the process, such as coalition partners, U.S. Marine Corps attachments, etc.

- Do not have tunnel vision and neglect to incorporate events from adjacent units into the wargame. Ensure information operations and political, military, economic, social, information, infrastructure, physical environment, and time (PMESII-PT) are incorporated. For example, refugees can disrupt movement schedules as surely as a minefield and newspaper headlines can change a campaign outcome.

The following are common issues with COA analysis:

- Failure to practice. From practice comes all MDMP understanding and proficiency. The field should not be the first time, or even the second or third, the collective staff is conducting a wargame together.

- Participants lack experience and education to provide useful or realistic input that is packaged clearly and succinctly. Participants do not understand their WfF's role in relation to other WfFs and "the big picture."

- Incomplete COAs: COAs must be complete or time will be wasted and confusion will prevail as COA analysis turns into COA development.

- Rarely is there enough time to wargame all COAs thoroughly and incorporate branches and sequels into the analysis. Choices must be made when planning the wargame (as in most planning steps) of what the priorities are and where planning risk can be tolerated.

- Wasting time during execution: The planning team leader must not be afraid to cut people off, use brief by exception, or shelve issues. It is COA analysis, not COA development or an open-ended brain-storming session. Avoid bogging down in minutia or arguments on battle damage assessment (BDA).

- Wargaming battle drills instead of COAs: COA analysis is not the time to rehearse or discuss unit or staff battle drills.

- The opposing force (OPFOR) changes the enemy COA substantially from previous briefings. For example, instead of wargaming against the most likely COA, which the friendly COA was designed to defeat, the wargame is against something the plan was not designed for, and that should be considered a branch.

- The OPFOR is not given the opportunity to fight back as a thinking and adaptive enemy.

- Units often focus planning, wargaming, and rehearsing only on "actions on the objective." However, units often encounter difficulties and culminate before reaching the objective. Ensure actions during contested movements or while forces are otherwise vulnerable, are adequately addressed.

COA analysis does the following:

- Creates shared understanding through visualization and discussion of friendly, enemy, environment, and other domains in time and space. Understanding informs decision making

- Refines existing decision points and identifies new ones, and identifies branches and gaps

- Facilitates identifying risks and mitigation to those risks, as well as opportunities to exploit

- Tests friendly COAs against various enemy COAs and variables, and illustrates possible outcomes, and facilitates identifying advantages and disadvantages of a COA

- Synchronizes and refines a COA across all WfFs

- Enables more detailed planning than in COA development. It allows for refining the plan, improving incorporation of enablers, and creating or updating products like synchronization matrices, situation templates (SITEMPS), graphics, tasks, and decision support matrices (DSMs). Detailed recordkeeping by all participants greatly eases the completion of the final order.

- Challenges assumptions and planning factors

- Facilitates the determining of information requirements

- Validates COAs as feasible, viable, and acceptable

COA analysis does not do the following:

- Predict with certainty: Wargames can predict plausibility, but there are too many variables to definitively predict probability.

- Create reproducible results: Expect neither a real operation nor a repeat of a wargame to unfold exactly as predicted.

- Address chance (i.e., black swans), or reveal the enemy's thought process

---

COURSE OF ACTION ANALYSIS METHODS
FROM FIELD MANUAL (FM) 6-0, *COMMANDER AND STAFF ORGANIZATION AND OPERATIONS*, 05 MAY 2014

"Three recommended wargaming methods exist: belt, avenue-in-depth, and box. Each method considers the area of interest and all enemy forces that can affect the outcome of the operation. Planners can use the methods separately, in combination, or modified for long-term operations dominated by stability."

The **belt method** divides the area of operations into belts (areas) running the width of the area of operations. The shape of each belt is based on the factors of mission, enemy, terrain and weather, troops and support available, time available, and civil considerations (METT-TC). The belt method works best when conducting offensive and defensive tasks on terrain divided into well-defined cross-compartments, during phased operations (such as gap crossings, air assaults, or airborne operations), or when the enemy is deployed in clearly defined belts or echelons. Belts can be adjacent to each other, or overlap.

This wargaming method is based on a sequential analysis of events in each belt. Commanders prefer the belt method because it focuses simultaneously on all forces affecting a particular event. A belt might include more than one critical event. Under time-constrained conditions, the Commander can use a modified belt method. The modified belt method divides the area of operations into three or less sequential belts. These belts are not necessarily adjacent or overlapping, but focus on the critical actions throughout the depth of the area of operations."

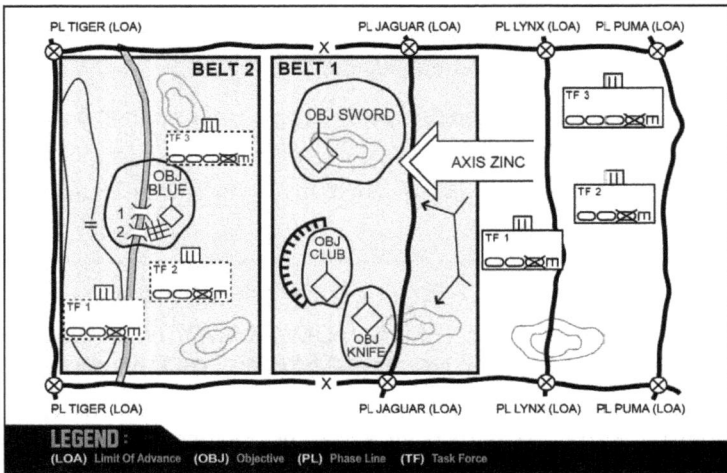

Figure 2-1. Sample Belt Method
(FM 6-0 page 9-28)

"The **avenue-in-depth method** focuses on one avenue of approach at a time, beginning with the decisive operation. This method is good for offensive COAs, or in the defense when canalizing terrain inhibits mutual support."

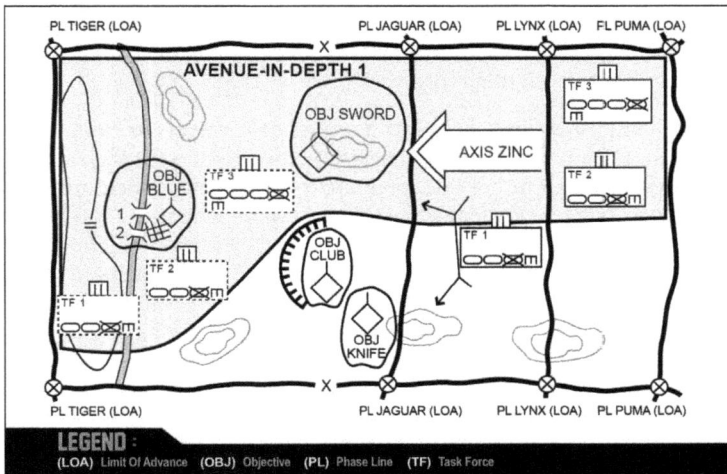

Figure 2-2. Sample avenue-in-depth method
(FM 6-0 page 9-29)

"The **box method** is a detailed analysis of a critical area, such as an engagement area, a wet gap crossing site, or a landing zone. It works best in a time-constrained environment, such as a hasty attack. The box method is particularly useful when planning operations in noncontiguous areas of operation. When using this method, the staff isolates the area and focuses on critical events in it. Staff members assume that friendly units can handle most situations in the area of operations, and focus their attention on essential tasks."

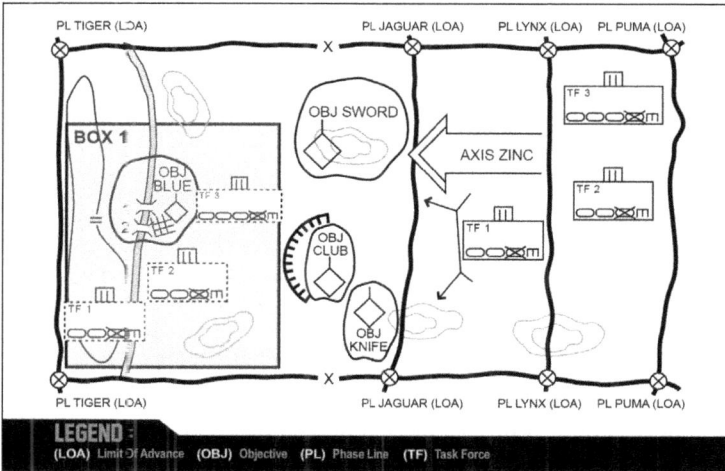

Figure 2-3. Sample box method
(FM 6-0 page 9-30)

Note: Staff must keep the inputs and outputs of the wargame in mind. The table below illustrates these inputs and outputs, and the process or checklist for setting up the wargame. Regardless of the method used, the approach is always action-reaction-counteraction.

➜ INPUT	PROCESS ⬇	OUTPUT ➜

• Updated running estimates
• Revised commander's planning guidance
• Course of action statements and sketches
• Updated assumptions
	• Gather the tools
	• List all friendly forces
	• List assumptions
	• List known critical events and decision points
	• Select the wargaming method
	• Select a technique to record and display results
	• Wargame the operation and assess the results
	• Conduct a wargame briefing (optional)
		• Refined courses of action
		• Potential decision points
		• Wargame results, such as:
		  – Decision support template and decision support matrix
		  – Synchronization matrix
		  – Potential branches and sequels
		  – Updated running estimates
		• Initial assessment measures
		• Updated assumptions

Figure 2-4. Input/output process (Source: FM 6-0[1])

During execution, the hard outputs, like DSTs, DSMs, and synchronization matrices, are important in assisting the commander with seeing and anticipating events and actions relating back to the wargame. These decisions may be planned, but as events in real-time occur, unanticipated decisions will occur. It is at this point that the critical skills of visualization will help reconcile the results of the wargame with the reality of the fight. What is changing? Why is it changing? What does it mean? Is this an opportunity or a crisis? How is the plan affected? What is the recommended reaction or counteraction?

The intangible result gained from the higher level of visualization (seeing your force, the enemy, and the terrain) and the wargame is that units will focus more on fighting the enemy than the plan. As changes occur during contact with the enemy, the staff's battle rhythm during execution should turn into an extension of the wargame, reactions and counteractions, with a greater understanding of realized threats and opportunities. In the end, enemy actions and unanticipated information can lead to the staff recommending new decisions or adjustments to the commander.

CONDUCT OF THE WARGAME
(ACTION-REACTION-COUNTERACTION)

> "Adversarial" is a key — perhaps the key — characteristic of wargaming. Wargaming is a competitive intellectual activity, and the primary challenge is usually provided by a combination of opposing players representing active, thinking, and adaptive adversaries and competitors.
>
> *Wargaming Handbook*
> UK Ministry of Defense, August 2017

Execution Steps

The actual conduct of a COA analysis wargame has the following steps:

- Conduct a COA analysis orientation pre-brief.

- For every COA and branch wargamed:

  o Brief the initial set for the COA about to be wargamed.

  o Play turns. For each turn:

    * Brief the turn overview.

    * Execute the turn: action-reaction-counteraction phases.

    * Adjudicate outcomes.

    * Conduct an end-of-turn assessment.

  o Conduct an end-of-COA/branch wargame review and assessment.

- Compare COAs.

The following are measures of wargame success:

- Can all participants visualize and understand how the plan will unfold?

- Are the plans being synchronized across all WfFs?

- Are the gaps, issues, and risks being identified and mitigated?

- Are advantages and disadvantages assessed against evaluation criteria?

- Is enough information being generated to complete a detailed plan?

  o Is information efficiently recorded?

  o Are there branches or events that require additional wargaming?

- Are identifying and developing decision points being emphasized? Decisions should:

  o Define what the alternative actions (options) are.

  o Define triggers and conditions (event, time, etc.).

  o Define information requirements to support the decision.

  o Define who is responsible for tracking the information requirements, and packaging them for the decision maker.

  o Define who makes the decision.

  o Synchronize the decisions/options across the WfFs.

WARGAME RULES AND METHODS

Russian Wargame of the Battle of Tannenberg

Just prior to World War 1, the Russian General Staff conducted a wargame of a Russian invasion of East Prussia by the First and Second Russian Armies in a two-pronged assault. During the wargame, each Russian army was isolated, and in turn destroyed, by the smaller, quicker German force. Furious, the two Russian commanders demanded the wargame's facilitator change his adjudication of how quickly the German forces could move, slowing them considerably. Under intense pressure, the facilitator restarted the wargame, slowing the Germans and allowing the two-pronged attack to succeed spectacularly. Four months later, when the same plan was executed on the real battlefield, the Germans moved as quickly as the facilitator had originally predicted and destroyed both Russian armies one at a time: The first one at Tannenberg and the second one a week later at Masurian Lakes. The Russians were ejected from East Prussia, suffering over 300,000 casualties.[2]

Games need rules.

A set of commonly understood rules, ideally codified in a standard operating procedure (SOP) and practiced routinely, greatly eases wargame execution, and allows participants to focus on analyzing their actions instead of focusing on how to play. Off-the-shelf wargames come with instructions, but the doctrine for COA analysis does not necessarily prescribe rules to conduct action, reaction, and counteraction. This leaves the staff on its own to determine the rules.

The Simulation.

The simulation is the engine that drives the adjudication and outcomes of decisions or planning. The simulation consists of tools to replicate the environment and actors (such as a map board and unit markers), rules for methods of decision making and action taking (such as the action-reaction-counteraction format), and methods for adjudicating the outcomes of decisions and actions (such as a correlation of forces and means (COFM) spreadsheet or loss table). A simulation may be extremely detailed and even computer-driven, such as Warfighting Simulation (WARSIM) supporting warfighter exercises, or it may be as simple as participants moving sticky notes around a sketch on a dry-erase board and applying their professional knowledge to create probable, realistic outcomes. Ultimately, the facilitator must manage the simulation and decide outcomes based on personal judgement and experience, incorporating input from the participants and other references. Simulations address variables such as:

• How to portray units and terrain, time, and distance

• Movement rates under various conditions

• Unit and column "footprints" and densities

• Combat outcomes and BDA

• Supply consumption

• The human dimension (morale, proficiency, fatigue, etc.)

• Detection, intelligence, surveillance, and reconnaissance

• Friction and fog of war

• How quickly subordinate units can react to orders or consolidate and reorganize

• Terrain and weather impacts

• Task performance such as entrench; breach an obstacle; establish a command post; and command, control, and communication (C3)

• Cyberspace electromagnetic activities (CEMA)

• PMESII-PT events

ADJUDICATION

Within the simulation, wargames require a method to assess the success or failure of a task, how long it takes to perform, and the associated losses. This is called adjudication.

Adjudication is a combination of art and science. It refines the battlefield calculus used during COA development. Relative combat power analysis and force ratio models, as well as relevant historical examples, can be used to inform estimated outcomes.

The following are rules for adjudication:

- Ensure a human is in the adjudication decision loop. This is important to identifying and incorporating variables that the simulation does not account for when determining likely outcomes, such as training proficiencies of teams and crews.

- Be consistent.

- Avoid favoring friendly forces.

- Use the most likely outcome as the basis for the COA analysis, not a less-likely outcome. Wargame less-likely outcomes as separate branches and possible decision points.

- If either the likely outcome or less-likely outcome is negative or less than desirable, refine the plan to mitigate and reduce risk. Always ask "What happens if we have higher than expected losses or less success at a task than expected? What is the risk and how can it be mitigated?"

- Utilize all staff and WfFs to identify and mitigate risks.

> **Tip.** If available, FA57 Simulation Officers, FA49 Operations Research and Systems Analysis Officers, and Red Team Officers can assist in adjudications.
>
> **Tip.** Oversized personalities can have oversized impacts on adjudication results. The facilitator must be forceful in keeping outcomes realistic, and not favoring those with the loudest voices or the most rank.
>
> **Tip.** Do not fight subordinate units' fights for them as part of the wargame.
>
> **Tip.** Do not turn the wargame into a battle drill rehearsal.
>
> **Tip.** Effectiveness of indirect fire is often overstated.
>
> **Tip.** Effectiveness of friendly forces (blue forces [BLUFOR]) weapons are often overstated.
>
> **Tip.** The human dimension is often overlooked; it is assumed Soldiers are immune to fear and fatigue.
>
> **Tip.** The Data and Analysis Center has software available on a classified network that can further assist in assessing the BDA.

Adjudication Methods

The choice of method is driven primarily by time available, resources, and staff experience. An experienced staff and facilitator may allow expedited adjudication in time-constrained environments. Inexperienced staffs must rely more heavily on consulting references and using calculators. The following methods can be combined for an optimal balance of efficiency and accuracy.

Talk Through. This is the simplest type of adjudication. The facilitator allows the participants to assess the outcomes of their actions, to include movement, task performance, and BDA. Other participants may question the stated outcome. The facilitator can then confirm or adjust if necessary. This allows the participants the opportunity to state the rationale behind their estimate, as they may have thought of something others did not. Conversely, if the facilitator vetoes the participants' estimates, that rationale should be given as well. This allows the facilitator to assess the personalities and experiences of the participants, and also affords the opportunity to educate and mentor participants in visualizing the battlefield. The quality of this method is completely dependent on the skill and experience of the staff and facilitator.

Analog Charts and Tables. Multiple references provide planning estimates for tasks based on historical data and experiments. These references can address everything from movements of units under various conditions to fuel and ammunition consumption rates to the time it takes to emplace a bridge. Simply find the appropriate table and read the results. Scale the task and unit size to the specific situation being evaluated. Finally, adjust the results for mission variables such as weather, visibility, fatigue, etc. Note that these references are just averages. Actual results will generally follow a bell curve, but may sometimes have a more randomized distribution.

Automated Calculators. Similar to analog charts and tables, several automated programs exist where the user can input data and get an estimate, such as time of travel tools. Again, the user must control for variables not addressed in the software.

CORRELATION OF FORCES AND MEANS

"The goal for using the calculator is not so much to predict the outcomes of engagements as it is to add some objectivity to the force allocation process and to facilitate staff synchronization of the warfighting functions, to achieve the effects directed in the plan. Rules of thumb for calculator shortfalls allow the staff to focus more on synchronization, by accepting the calculator outcomes as good enough rather than an intellectual tug of war between the S2 and S3 over whether a system or unit was truly destroyed. Wargaming will progress more smoothly, making the outcomes more timely and synchronized."

Dale Spurlin and Matthew Green
Demystifying the Correlation of Forces Calculator, Infantry Magazine
Jan-Mar 2017

COFM is the nickname for a mathematical model that allows comparing various forces' relative combat power and effectiveness in specific mission sets against each other, and coming up with objective and "scientific" estimated losses and kills. It translates each unit into a "force equivalent (FE)" rating, that quantifies the unit's combat abilities.

COFM and relative combat power assessments use the historical force ratio table (see Table 2-1). These ratios are considered the ratios where success is a 50:50 proposition. Additional combat power is required to create a true advantage for one side or the other.

Table 2-1. Historical minimum planning ratios (Source: FM 6-0[3])

Friendly Mission	Position	Friendly:Enemy
Delay		1:6
Defend	Prepared or fortified	1:3
Defend	Hasty	1:2.5
Attack	Prepared or fortified	3:1
Attack	Hasty	2.5:1
Counterattack	Flank	1:1

Tip. Each staff officer should actively collect useful tables and calculators for their smart books. Among others, the Army Techniques Publicaticn (ATP) 5-0.2, *Staff Reference Guide,* to be posted in 2020, has many useful references as a starting point.

CAUTION: The FE rating typically reflects the unit's inherent effectiveness at engaging in combat the way the unit was designed. It does not account for how suited a unit's predominant weapon systems are against specific target types or circumstances. For example, an artillery unit's FE is based on its indirect fire abilities, not how well it would perform in a direct fire engagement against a tank unit.

CAUTION: It may be tempting to "pile on" units in an engagement to achieve a higher relative combat power ratio, but only include those units that could realistically directly participate given time and space considerations. The law of diminishing returns applies here, and there is the risk of creating a dense target rich environment.

CAUTION: There is no proponent tasked to update the COFM to account for changes to organization, technology, etc. Information may be stale.

The facilitator must adjust COFM FEs to reflect significant variables. COFM does not account for terrain, training, morale, C3 degradation, weapon ranges, or other asymmetries. It does have a limited ability to account for the degree of fortification. Variables can be accounted for in several ways: increasing or decreasing the unit strength to reflect an advantage or disadvantage in a given situation, manually adjusting the outcome, or if using an analog chart, shifting a column, which changes the force ratio.

Digital COFM. The Department of Tactics at the Command and General Staff College (CGSC) created a COFM calculator in an excel spreadsheet. It uses data derived from studies at the U.S. Army Training and Doctrine Command (TRADOC) G-2/7 (formerly known as TRADOC Intelligence Support Activity [TRISA]) and TRADOC Analysis Center (TRAC), and is used as a tool within the school house.

(**Note:** TRAC is now assigned to Army Futures Command's Futures and Concepts Center.) The examples on pages 58 and 59 are derived from that calculator.

The procedure for use is simple: From a drop-down menu, select all friendly forces participating and input their strength percentages (default is 100 percent), the number of this type of unit, and their mission type. Do likewise for enemy forces. The COFM calculator will provide a casualty estimate. Ensure the facilitator adjusts for variables. This can be done by increasing or decreasing units' percentage strength within the calculator, or by adjusting the final result manually. It is also possible to have fractions of units instead of whole numbers. The authors of the tool recommend using lower-echelon combat units, such as battalions instead of brigades, as this more accurately reflects what is actually engaged in a given fight. For higher echelons and longer duration fights, brigades can be used to help reflect the resiliency and sustainability of those units.

Analog COFM. The following is "a way" to manually calculate the COFM using simple math: The FE numbers are scaled by one fourth, and then rounded to the nearest whole number to make them more manageable. For example, a full-strength, armor-heavy M1A2 combined arms battalion has a nominal FE of 37.24. This is an awkward number to manipulate, so their scaled FE is 9. To calculate the COFM, add up all the scaled FEs on each side and find the ratio. Consult the damage table for the appropriate missions (for example: deliberate attack versus hasty defense) and read the losses for each side. If in the aggregate, there are other variables that have a significant impact on combat, favoring one side against the other, shift the column left or right, which effectively changes the force ratio used to calculate the outcome.

ANALOG CORRELATION OF FORCES AND MEANS AND STEP LOSSES

A simple method to track damage is through step losses with a damage table. Each step loss is an abstract measurement of combat power and corresponds to a transition in the military shorthand of "green, amber, red, black." (See Figure 2-5 for corresponding percentages.) While not precise, this is a quick way to track units' combat effectiveness. Unit icons can even have the scaled FE for each step loss marked directly on them. Write directly on the icon to indicate the current FE and green, amber, red, or black status. Alternatively, rotate the counter so the current status is always toward the top of the game board. (See Figure 2-6 for an example of a customized board game icon.)

STEP LOSSES:

1 15-30%

2 35-50%

3 55-70%

E >75%

GREEN [COMBAT CAPABLE]
INDICATES: UNIT IS AT 85% OR > STRENGTH

AMBER [COMBAT CAPABLE MINOR DEFICIENCIES]
INDICATES: UNIT IS AT 70-84% STRENGTH

RED [COMBAT INEFFECTIVE MAJOR LOSSES/DEFICIENCIES]
INDICATES: UNIT IS AT 50-69% STRENGTH

BLACK [REQUIRES RECONSTITUTION BEFORE NEXT MISSION]
INDICATES: UNIT IS AT < 50% STRENGTH

E [NO LONGER EXISTS AS A FUNCTIONAL FORMATION]
INDICATES: UNIT NO LONGER FUNCTIONAL

Figure 2-5. Example of friendly to enemy loss percentages in green, amber, red, and black statuses. (Source: CGSC Tactics Division)

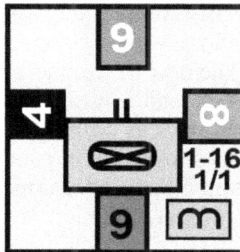

Figure 2-6. Example of a customized board game icon. (Source: CGSC Tactics Division)

Tip. Units still suffer losses in combat, even if those losses are not large enough to warrant a step loss. A unit that participates in multiple combats but does not get a step loss on the damage table might eventually accrue a step loss based on the facilitator's and other participants' judgement.

Figure 2-7 is an example of an analog unit FE matrix from the COFM generator, which is the simplest method to track damage through step losses. It has scaled step losses and a combined arms battalion marker with "green, amber, red, black" and corresponding FEs for each step loss.

Type	Base	Green	Amber	Red	Black	Type	Base	Green	Amber	Red	Black
FA Bn (M109A6)	28.13	7	6	4	3	Infantry Bn (BMP-3)	41.12	10	8	6	4
AR Combined Arms Bn (M1A2)	37.24	9	7	6	4	Tank Bn (T-80U)	28.96	7	6	4	3
IN Combined Arms Bn (M1A1)	37.59	9	8	6	4	FA Bn (2S19)	21.42	5	4	3	2

LEGEND (AR) ARMOR (BN) BATTALION (FA) FIELD ARTILLERY (IN) INFANTRY

Figure 2-7. Example of an analog unit friendly-to-enemy matrix. (Source: CGSC Tactics Division)

FRIENDLY to ENEMY FORCE RATIO	1:6	1:4	1:3	1:2	1:1	2:1	3:1	4:1	5:1
FRIENDLY vs ENEMY	0.200	0.250	0.333	0.500	1.00	2.00	3.00	4.00	5.00
PERCENTAGE of FRIENDLY FORCES	0.17	0.20	0.25	0.33	0.50	0.67	0.75	0.80	0.83
Deliberate Attack vs Hasty Defense	E	E 1	3 1	3 1	1 2	1 2	1 2	3	3
Deliberate Attack vs Deliberate Defense	E	E	E	3	2 1	1 1	1 1	2 1	3
Hasty Attack vs Hasty Defense	3	3 1	3 1	3 1	2 1	1 2	1 2	1 2	2
Meeting Engagement vs Meeting Engagement	3	3	2	2	1	1	2	3	E
Hasty Attack vs Deliberate Defense	E	E	E	E	3	2 1	1 1	1 1	1

Figure 2-8. Example of further friendly-to-enemy ratios. (Source: CGSC Tactics Division)

Example of an analog scaled COFM with step loss

Two green M1A2 battalions (FE 9), supported by an amber FA battalion (FE 6), conduct a hasty attack against a green BMP-3 battalion (FE 10) conducting a deliberate defense. The total friendly FE is 24 (9+9+6) against an enemy FE of 10. The ratio is 2.4:1, which is then rounded to 2:1. Friendly forces lose two step losses for each unit, and the BMP loses one (no losses for the artillery). The new FE for the friendly tank units is now six and in a red status. The BMP unit is now FE 8 in an amber status. This does not account for variables, such as training or visibility. Assess if all the variables in total would have a substantial impact. For example, consider the .4 from the 2.4:1 force ratio that rounded down to a 2:1. If there was a substantial impact, shift the column to the left or right to assess losses.

CONDUCTING ACTION-REACTION-COUNTERACTION CYCLES

The basis of the wargame is the action-reaction-counteraction cycle. The facilitator drives this process, ensuring timelines are met and the staff is focused. He avoids getting mired or going off on tangents, and produces quality outputs. The following is an example of how the cycle works, which assumes BLUFOR have the initiative:

• Determine initiative, which then determines which force initiates the turn with the action, and which side reacts. Normally the offensive side starts with the initiative. There are times when there will be a change of which force (friendly or enemy) initiates the action-reaction-counteraction sequence.

• The facilitator utilizes the synchronization matrix to call on WfF briefers in a logical order.

• The WfF briefers in turn describe the locations and actions/activities in their respective WfF. Briefers are concise, sharing only what is necessary for the recorder and scribe to capture and to create shared understanding with the other participants.

• The level of success or failure of the action is assessed as per the adjudication method chosen by the facilitator.

• Continue through all WfFs and briefers, until all have had the opportunity to provide input. Not every WfF will always have input.

• The opposing forces (OPFOR) conduct their reaction phase to the BLUFOR's actions, and address all relevant WfF activities. In some cases, with the facilitator's permission, the OPFOR may interject a reaction at an earlier point before all BLUFOR actions are taken. The facilitator adjudicates the reaction, including BDA, as required.

• A counteraction may then be initiated against the OPFOR reaction. Again, the facilitator adjudicates as required.

• Additional iterations of reactions-counteractions can take place within the same turn, following a natural flow of discussion. However, it should be manageable and the facilitator should decide if breaking this turn into multiple turns is beneficial.

• If participants identify risks, gaps, requests for information (RFIs), advantages/disadvantages, issues, etc., during the turn cycle, they take note and either bring it up immediately or brief it during the end of turn assessment.

- The scribe and recorder enter data into the synchronization matrix and other documents as necessary.

- All participants take appropriate notes, and update products and staff estimates. At turn end, the facilitator leads the end of turn assessment. This discussion identifies risks, advantages, disadvantages, RFIs, issues, branches, etc., associated with this turn.

Tip. The recorder projects a synchronization matrix where all can see it, and fills it out as the wargame progresses.

Tip. The facilitator may wait until all actions-reactions-counteractions are complete before adjudicating outcomes, particularly combat outcomes.

Tip. The scribe records advantages, disadvantages, risks, opportunities, assumptions, RFI decisions, and areas or contingencies needing further study as they are identified.

Tip. Participants update their planning products for this COA as they progress through the wargame.

Conducting Turns Using an Operational Framework

For each turn, it may be useful to divide the action into operational frameworks and wargame each part of the framework individually, from deep, close, and consolidation/security/rear/sustainment. The OPFOR mirrors the BLUFOR.

Wargame Discussion Questions

The purpose of the game is to explore the COA, not just to check the blocks on a synchronization matrix. The facilitator and other participants should ask the following questions throughout the course of the wargame to promote discussion and thorough analysis:

- What is the risk at this point to this unit? To the mission? How can it be mitigated?

- What happens if losses during an event are higher than expected, or if there is difficulty completing a task or movement?

- What other assets or combat multipliers can be used to facilitate a task or mitigate a risk?

- Does the battlefield array and architecture balance force protection and rapid action and massing?

- What are the weather, light, and terrain impacts? What is the key terrain?

- What is the enemy commander thinking at this point? What is his next decision?

- What are the enemy reserves and triggers for commitment?

- What combat multipliers might the enemy bring into play?

- How does something impact decisions? What new information requirements does this generate?

- Where are casualties right now? Where are the supply convoys (coming and going)?

- What is needed to avoid culmination?

- What events in the area of interest have an impact and need to be tracked? How does higher headquarters see the fight at this point?

> **Tip.** The Maneuver and OPFOR leads should routinely ask each other (and themselves) questions along the lines of "What could I do to interfere with the execution of the desired action?" This assists to identify risks, mitigations, decision points, branches, and opportunities.

Turn Completion

The facilitator conducts an end-of-turn review and assessment:

- Confirm the COA still meets the criteria of feasible, acceptable, and suitable, and meets all the commander's screening criteria.

- Review changes to:

  o Decision points and any substantial adjustments to the COA

  o Assumptions, task organization, taskings, timings, additional guidance, coordination required, control measures, information requirements, and RFIs generated

  o Gaps and "tabled" or "parking lot" topics

  o Branches or issues that require further attention.

- Conduct an assessment by evaluating the WfF against the established criteria as applied during that turn.

  o Advantages

  o Disadvantages

  o Risks

• The facilitator identifies and prioritizes remaining branches to be wargamed. The staff wargames the desired branch (before going to another COA if possible), and conducts an additional end-of-turn assessment.

> **Tip.** Add rows for evaluation criteria and risk, and add notes to the bottom of the synchronization matrix to help record results.

WARGAME VIGNETTE

> "If you take a flat map
> And move wooden blocks upon it strategically,
> The thing looks well, the blocks behave as they should.
> The science of war is moving live men like blocks.
> And getting the blocks into place at a fixed moment.
> But it takes time to mold your men into blocks
> And flat maps turn into country where, creeks and gullies
> Hamper your wooden squares. They stick in the brush,
> They are tired and rest, they straggle after ripe blackberries
> And you cannot lift them up in your hand and move them.
> It is all so clear in the maps, so clear in the mind,
> But the orders are slow, the men in the blocks are slow
> To move, when they start they take too long on the way
> The General loses his stars, and the block–men die
> In unstrategic defiance of martial law
> Because still used to just being men, not block parts."
>
> Stephen Vincent Benét
> *John Brown's Body* (1928)

The following vignette covers one turn in the middle of an armored brigade combat team (ABCT) offensive wargame on an analog map board, using the belt method. Several turns have already happened. The scheme of maneuver is for Task Force (TF) 1, Shaping Operation, to fix the enemy on Objective (OBJ) Club, to prevent repositioning to OBJ Sword. TF3, Decisive Operation, seizes OBJ Sword to protect the flank of an adjacent brigade. TF2 is the reserve.

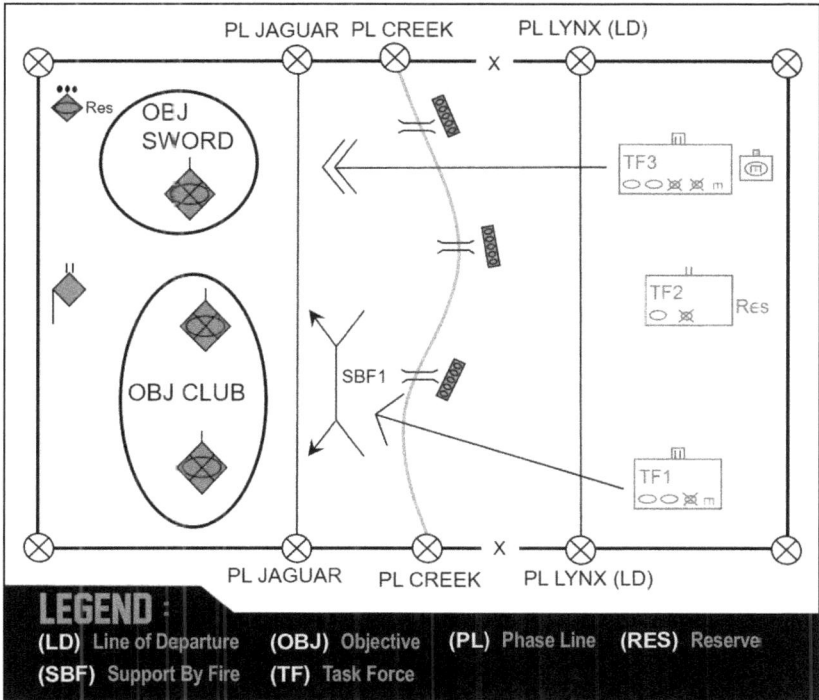

Figure 2-9. Wargame vignette graphic (deliberate attack)

Turn Overview

Facilitator. The next turn is Phase 2A: Movement to Phase Line (PL) Jaguar, and isolation of OBJ Sword from hour (H) +3 to H+4. Okay, what is the larger picture area of interest update?

Intelligence. No weather or light changes. Low ceilings and intermittent rain continue.

Maneuver. Enemy long-range fires have been degraded by 50 percent. The Division has just executed division Decision Point 2, which is for our brigade and the 4th Brigade to cross PL Lynx and simultaneously continue to attack in-zone. The 2nd Brigade remains in its tactical assembly area (TAA).

OPFOR. I am still in the defense. My security zone elements that were forced to withdraw are reconsolidating west of PL Tiger. I haven't committed any reserves from battalion through division. My next decision is where to commit the battalion and regimental reserves.

Action Phase

Intelligence. Shadow 1 and TF1 scouts observe Named Area of Interest (NAI) 1 on OBJ Club, and refine the target group, task group (TG) 11. Shadow 2 observes Target Area of Interest (TAI) 2 for the enemy's regimental reserve.

Maneuver. All units are still green. TF1 crosses the line of departure (LD) and attacks in-zone to occupy Support by Fire (SBF) 1, in order to fix enemy forces on OBJ Club to prevent repositioning of forces against TF3's attack on OBJ Sword. On order, TF3 attacks in-zone to seize OBJ Sword, to protect 4th Brigade's flank. TF2 has no change, still in reserve.

Decision points. Initiating TF3's movement to attack OBJ Sword is Decision Point 1 for the brigade.

Aviation. Air Weapons Team (AWT) 1 supports TF1. We'll have one hour time on-station for the Apaches until the forward arming and refueling point (FARP) moves to PL Puma, which doesn't happen until after OBJ Sword is secured. AWT 2 is on-ground at the FARP, and is 15 minutes from PL Lynx and 25 minutes from PL Jaguar.

Fires. Fire support coordination line (FSCL) is PL Tiger. Field artillery (FA) is in the Position Area for Artillery (PAA) 2. We are down to two battalion 6's of dual-purpose improved conventional munition (DPICM). Priority of fires (POF) to TF1, and shifts to TF3 when TF3 crosses PL Lynx. We'll shoot Target AB0001, suspected enemy battalion headquarters (HQ), when TF1 crosses the LD and TG 11 is on-order. Q36 establishes critical friendly zones (CFZ) on ABF 1. Division artillery (DIVARTY) is providing counter battery.

Intelligence. If electronic warfare (EW) jams with the fire mission on the CP, it would be a good chance to do some collecting. Their communications nets will be very active trying to figure out what's going on.

Air liaison officer (ALO). We have no apportioned missions. The division has six on-call close air support (CAS) missions we can request but priority is to 4th Brigade, and targeting the enemy division reserve once identified.

Protection. The team engineer follows TF3, and has to be prepared to conduct an in-stride breach on OBJ Sword. Enemy prisoner of war (EPW) Collection Point 1 has been established.

Air defense officer (ADO). ADO sections are opconned (operationally controlled) for movement with the TFs. Sentinel and one section are set on PL Lynx, and can cover up to PL Jaguar.

Chemical, biological, radiological, and nuclear (CBRN).
Decontamination (decon) platoon follows TF1 and moves to CP 1 on
PL Lynx. We have enough supplies to do thorough decon of two tracked
companies.

Sustainment. Class V resupply for artillery departs from the brigade
support area (BSA). Estimated time of arrival (ETA) H+6 vicinity PAA-3.
Brigade ambulance exchange points (AXPs) have been established at CP3
after TF1 LDs, and CP4 after TF3 LDs.

Command and control. The tactical command post (TAC) follows TF1.

Civil military officer (CMO). Occasional internally displaced persons
(IDPs) are in sector. Not enough to slow movement, but enough to be
careful about positive identification (ID) of targets.

Reaction Phase

OPFOR. I have my raven equivalents and dismounted scouts screening all
avenues of approach, and I detect TF1 crossing the LD. The rain has flooded
the creek and made crossing anywhere but a bridge difficult, and I have
thoroughly mined the bridges and easy fording sites.

OPFOR. My radar picks up your artillery shooting at my command post,
and my DIVARTY shoots a BM21 rocket counter-battery. Then I engage
TF1 with my regimental artillery group (RAG) when they attempt to cross
the creek.

OPFOR. As TF1 sets its ABF, I engage with anti-tank guided missiles
(ATGMs) and mortars.

Maneuver. We just destroyed your command post with artillery. How are
you calling in indirect fire (IDF) when we cross the creek? Your forces don't
have C3, and you are confused and demoralized.

OPFOR. You shot my CP an hour ago when you crossed the LD. I've had
more than enough time to reestablish C3.

Maneuver. Okay, we need to adjust the timing of engaging the CP and
when we prep OBJ Club. OPFOR, when would be the most disruptive time
to hit your CP? We'll adjust the trigger.

OPFOR. About five minutes before you get in direct fire range.

OPFOR. And I blew up ten of your tanks with my artillery crossing the
creek, and another 15 with my ATGMs when you moved into the ABF.

OPFOR. And my counter battery destroyed six artillery pieces.

Facilitator. Hold up there on the BDA for now.

Facilitator. OPFOR, what decision points do you have?

OPFOR. My next decision is to commit my battalion reserve, but the trigger for that isn't until I identify your main effort crossing PL Jaguar.

Counter-Action Phase

Intelligence. None

Maneuver. TF1 employs counter-unmanned aircraft systems (UASs). TF1 emplaces scissor bridges to rapidly cross the creek to avoid artillery fire and mines, and attacks to destroy enemy intelligence, surveillance, and reconnaissance assets in-zone en route to ABF 1. Under cover of fires, TF1 occupies ABF 1 and engages enemy on OBJ Club

Fires. DIVARTY shoots counter battery against the RAG, targeting TF1's creek crossing as acquired. On-order, our artillery shoots TG11 on the objective. We are now down to one battalion 6 of DPICM.

Protection, sustainment, command and control. No change.

Facilitator. Okay, let's do some BDA. Let's start with IDF. First, I assess the enemy battalion CP is now at 50 percent effectiveness, and can't conduct any C2 for thirty minutes. Next, let's address the enemy's 122mm RAG shooting at the creek crossing. Fires, how long will the enemy be able to shoot before we can get effective counterbattery on them, and how much can they shoot before then?

Fires. It's already covered in a CFZ, so maybe four minutes from acquisition to splash. The enemy CP is degraded from our fire mission. We'll say they can shoot about 75 rounds at each of the two crossing sites. The target is spread out and moving. According to my artillery effects table, we'll say one scissor bridge and three tanks are destroyed. The counter battery will destroy three enemy systems and force them to move, taking them out of the fight until they reposition.

Maneuver. That's a loss of four percent for TF1, which brings them to 90 percent. They're still green. What are the effects on OBJ Club?

Fires. Let's see. A battalion 6 of DPICM against entrenched vehicles spread out. I'd say four BMPs and one tank. Artillery will conduct survivability moves by battery to provide continuous suppression of the OBJ, but will only be shooting high explosives (HE) for suppression at this point. We can shoot for about 15 minutes of suppression.

Facilitator. Okay, let's put the friendly versus enemy maneuvers in the COFM spreadsheet. This will be a friendly deliberate attack versus an enemy deliberate defense. Maneuver, OPFOR, what's your combat power?

Maneuver. One tank-heavy combined arms battalion at 90 percent strength, and one paladin battalion at 90 percent strength.

OPFOR. After the artillery prep, I've got two BMP-3 companies at 80 percent in fortified positions.

Facilitator. Okay, who else can bring something to this?

Aviation. They've got the AWT 1, too.

EW. We can attempt jamming as well, but we can't do it for long.

(The COFM operator selects an M1A2 battalion at 90 percent, a Paladin battalion at 90 percent, and an Apache company at 25 percent. There is no "AWT" or platoon selections available so the unit is scaled accordingly. A BMP-3 battalion is selected at 80 percent, but under "Number," a ".7" is used instead of a "1" because only two of three companies from the battalion will be engaged per the SITTEMP for the objectives.)

Facilitator. What else impacts this? Subtract 10 percent from the enemy strength for the jamming and C3 disruption.

Facilitator. We've got a force ratio of about 3 (2.98):1, and losses of 21 percent versus 34 percent. Any other input? That puts TF1 at 69 percent effectiveness, which is amber, and the BMP companies to 46 percent, which is black. TF1 has accomplished its task of fixing.

Friendly Forces					Enemy Forces				
Number	Strength	Type	F.E.	Total	Number	Strength	Type	F.E.	Total
1	25%	AVN Co (AH-64)	5.24	1.31	0.7	70%	Infantry Bn (BMP-3)	41.12	20.15
1	90%	FA Bn (M109A6)	28.13	25.32		100%			
1	80%	Combec Arms Bn (M1A2)	37.24	33.52		100%			
	10%					100%			
	10%					100%			
	10%					100%			
	100%					100%			
	100%					100%			
	100%					100%			
	100%					100%			
	100%					100%			
	100%					100%			
	100%					100%			
Friendly Force Equivalent			60.14		Enemy Force Equivalent			20.15	
Ratio of Friendly to Enemy					Ratio of Enemy to Friendly				
2.98:1					0.34:1				
Deliberate Attack		<- Mission ->			Deliberate Defense				
21%		<- Est. Losses ->			34%				

LEGEND: (AR) Armored (AVN) Aviation (Bn) Battalion (CO) Company (FA) Field Artillery (F.E.) Friendly to Enemy Ratio

Figure 2-10. Example of a force ratio calculator matrix for the vignette (Source: CGSC Tactics Division)

Facilitator. What happens if TF1 takes heavy casualties before it can make it into ABF 1? Anything else we can do to mitigate that from happening, and if it does happen, what next?

Maneuver. Because TF1 is just a fixing force, it doesn't take as much combat power to just fix. The risk is low of us not being able to get enough force across to fix.

Facilitator. How long until TF3 gets up?

Maneuver. That's Decision Point 1. (Reviews the decision point worksheet.)

Figure 2-11. Decision Support Matrix

Facilitator. Is Decision Point 1 still valid and complete? Okay.

Turn Completion

Facilitator. Any gaps, issues, or changes to assumptions? Alright, is it still feasible, suitable, and acceptable? Good. Let's discuss evaluation criteria and risk. The evaluation criterion we're using are tempo and flexibility.

Maneuver. Tempo is a disadvantage right now. We have TF1 on the SBF for too long, and they can bleed out before TF3 gets there. I'd classify that as a risk. We can mitigate that by having TF3 LD earlier and hold up closer to OBJ Sword.

Facilitator. How long will it take TF3 to get to the objective?

Maneuver. TF3 has about a 15 kilometer cross country movement over slow-go terrain in tactical formation. (Inputs data into route planning tool.) It will take about an hour for them to reach PL Jaguar and begin the attack on the main objective. That's against light resistance. Due to the rain, we'll add another 15 minutes. Engineer, how long to emplace two scissor bridges and cross five tracked companies?

Engineer. (Consults planning tables) About 30 minutes if things go smooth.

Maneuver. Okay, let's have TF3 LD about 45 minutes after TF1, around when TF1 reaches the creek. Create an attack position graphic for TF3 to hold up into until ordered to continue to attack. We'll update Decision Point 1 to reflect this.

Manuever. Flexibility is an advantage. We have a very robust reserve and TF1 can still accomplish their task even with heavy casualties.

Fires. Culmination is a disadvantage and a risk. If the field artillery (FA) shoots much more than expected, we'll be black on ammunition, and have to stop.

Sustainment. Maybe we can mitigate the Class V by flying in some of the resupply, instead of trucking it all the way from the port to the BSA.

OPFOR. You have a planning gap or risk from my perspective. You don't have a plan to locate and destroy my regimental reserve tank company.

Engineer. I've got a risk. We've only got four scissor bridges. If we start losing them, the brigade can still cross the creek, but it will be very slow. I recommend we send a request or RFI for additional bridge resources.

Facilitator. Okay, let's make those adjustments. (Has the scribe review pertinent findings and sets the next turn.)

This vignette example illustrates the interactions and frictions that take place in a good COA analysis (wargaming), where the facilitator, the executive officer or G-5 adjudicates actions and reviews decision points as the staff visualizes the fight in time and space.

Endnotes

1. Field Manual (FM) 6-0. *Commander and Staff Organization and Operations.* 05 MAY 2014.

2. Kretchik, W. E. (1991). *The Manual Wargaming Process: Does our Current Methodology Give Us the Optimum Solutions? A Monograph for the School of Advanced Military Studies.* Fort Leavenworth, Kansas: School of Advanced Military Studies.

3. FM 6-0. Table 9-2

CHAPTER 3

Thoughts on Training the Staff

"You can ask me for anything you like, except time."

Napoleon Bonaparte
The Corsican: A Diary of Napoleon's Life in His Own Words

There is never enough time, or at least that is the perception. Training begins with a solid home-station training plan. Although the commander is responsible for training his staff, he has his executive officer (XO) or senior planners to assist with this training and with devising a home-station training plan. Staff training does not happen without command emphasis.

Staff tasks are like any other unit collective or individual task: Proficiency requires education, training, and practice. Without deliberate effort and command support, results will be haphazard at best. The commander and his primary staff officers should develop yearly nested training plans that address all aspects of staff training and education. These events should be planned, resourced, briefed, and approved during the quarterly training brief, added to calendars, and given as much protection as any other training event that builds readiness. Staff members must also actively take responsibility for their own professional development and training, as well as the training for any subordinates they may lead.

The goal is to create staff members who have an intuitive understanding and visualization of both their warfighting function (WfF) and the battlefield. In addition, staff members must understand their role during course of action (COA) analysis, as well as what to provide and how they add value. Based on observations from combat training centers (CTCs) and other exercises, the following are key trends staffs must address in their training:

• Lack of individual technical competencies. Wargame participants must be subject matter experts on their WfF and branch. For example, an engineer officer must speak authoritatively on terrain and the composition, capabilities, and employment of different bridging companies. Participants must visualize time, space, resources, and event outcomes within their areas of expertise. Participants must have access to, and be familiar with, reference manuals. They must study and conduct professional development at every opportunity. This also includes competency of reverse WfFs, understanding their WfF's threat, and coalition counterparts.

- Failure to practice. Repetition, repetition, repetition! The military decisionmaking process (MDMP) steps, such as COA analysis, that are not routinely trained and practiced, invariably lead to poor outcomes and take more time to accomplish. Make this a battle drill so mental energy is spent on analyzing the plan, not on figuring out how to execute the wargame. Units must practice often and include all planning enablers, including attachments, even if they are not permanently assigned to the planning team. Ensure the associated briefings are practiced as well.

- Failure to understand how the unit fights and commanders think. All units and individuals have personalities, cultures, etiquettes, pet peeves, etc. Planners must understand and visualize how these variables will affect the fight. This begins with understanding doctrine and unit standard operating procedures (SOPs), but does not stop there. How does the commander like to conduct counter reconnaissance and employ reserves? How are command and control nodes used? How does the commander like to use attack aviation? What are the personality traits of higher headquarters (HHQs), adjacent units, staff, and subordinate units that will affect planning and execution? What is the training level of each subordinate unit? Officer professional development allows commanders and key staff to communicate their visions. Staff exercises, team building, and even playing commercial wargames can help with this visualization.

The following are tools to address focus areas:

- Professional development sessions. Examples include discussions of professional readings, common operating environment updates, capability briefs, military history presentations, and how-to instruction on tasks such as gap crossings.

- Staff rides. These historical on-site studies are helpful, if the time and the resources are available.

- Staff exercises. These are either self-resourced or nested. They can range from table top exercises (TTXs), such as a simple tactical exercise without troops, to simulation-driven command post exercises with deployed command posts and 24-hour operations. Many units fail to take advantage of home-station events, such as gunnery, as opportunities to train the staff.

- Professional reading. This can be mixes of history books, professional journals, and publications. Many books without an obvious military connection are also useful and should not be discounted.

- SOP working groups. It is useful to periodically walk through and discuss planning procedures and products to create understanding and ensure products are current and distributed. It can be particularly effective conducting these in support of after action reviews (AARs) after major events.

- Orders. Keep and use old unit orders or reach out to the CTC, Mission Command Training Program, or the Command and General Staff College for old orders, which can be used as a driver for MDMP training.

- Administrative missions and taskings. Use the MDMP and orders publications for non-tactical events. Using the MDMP familiarizes the staff with tools and formats. This allows focusing on the output, and not the procedure. Although this may not always be a perfect fit, everyday garrison planning activities can provide training opportunities.

Breaking the MDMP process down into its parts aids in the development of training plans that seek to improve one step at a time. A new staff trying to take on the entire process at one time can lead to frustration and the development of bad habits. Focusing on one aspect of the MDMP at a time will allow deeper understanding and developing better individual and staff collective habits. Keep in mind that the output of one step is needed to train the next step. Further, repetition will allow the staff to attain a battle rhythm and level of efficiency for planning high-tempo large-scale combat operations. Figure 3-1 on the following page breaks the MDMP into its logical parts. Next, develop a training plan for Step 4, COA analysis (wargaming).

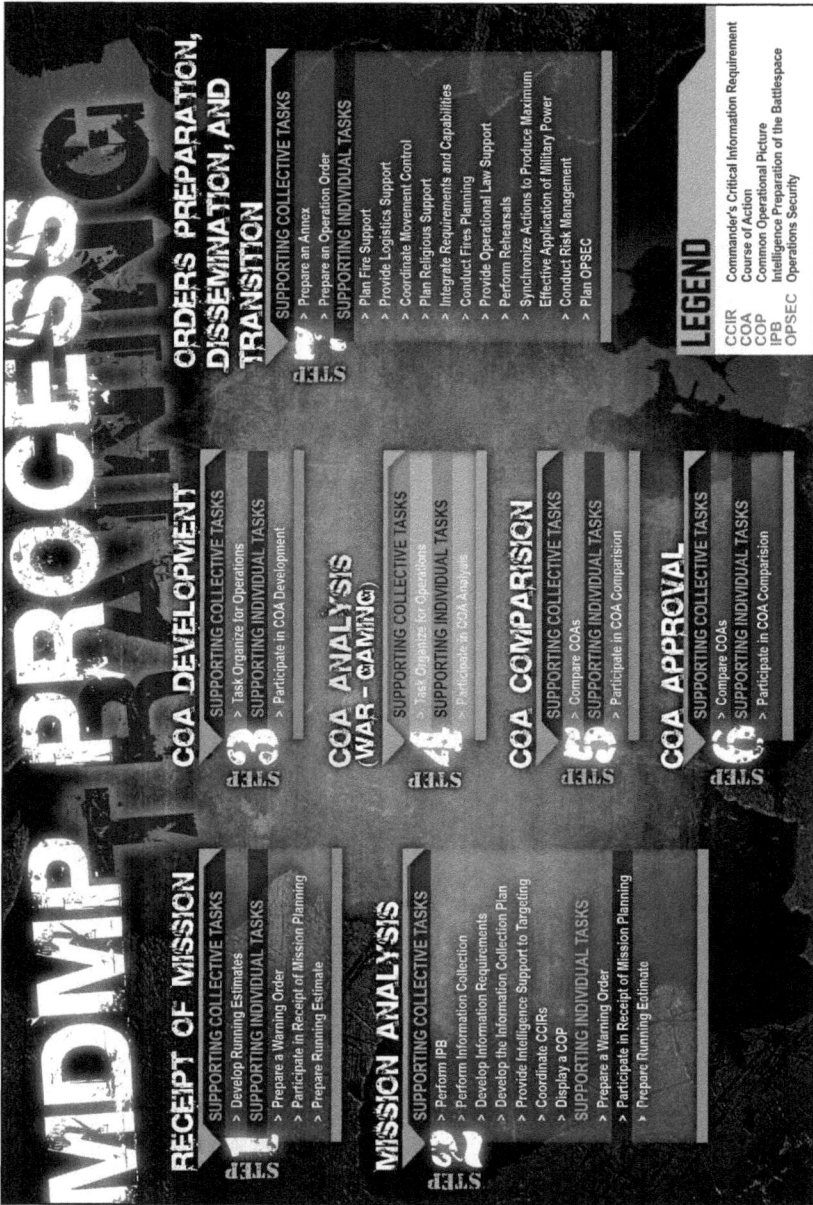

Figure 3-1. Steps of the MDMP

EXAMPLE OF A STAFF TRAINING PLAN

Below is a simple example of a training concept using what has been discussed to develop the conative skill of visualization and improve COA analysis. **Note:** Appendix A on page 73 provides a comprehensive reference for COA analysis (wargaming) tasks, conditions, and standards with detailed performance measures and outputs.

With the commander's guidance specific to the unit's mission essential task list (METL) and focusing on COA analysis (wargaming), the XO developed the following training plan concept. The training event will take place over a seven-week period and will culminate with practicing several iterations of Step 4 of the MDMP, COA analysis, with the main command post (CP) deployed to the field during platoon situational training exercise (STX) evaluations.

The training objective is to improve staff ability to conduct COA analysis. The terminal learning objective is to improve COA analysis, and the enabling leaning objective is to improve visualization (seeing ourselves, the enemy, and the terrain).

Using the crawl, walk, run approach, the staff will execute leader training, consisting of leader development classes, practical exercises using wargames, and TTXs. In this example, the XO initially uses chess as a means to develop individual conative skills and then uses the board game Kreigsspiel (mentioned in Chapter 1 of this handbook) to develop staff collective conative skills. The TTXs will further develop visualization of seeing ourselves, seeing the terrain, and seeing the enemy. This training event will culminate with conducting at least three full wargames in a field environment. Figure 3-2 is a graphic representation of this major staff training event.

WEEK / OBJECTIVES	MONDAY	TUESDAY	WEDNESDAY	THURSDAY	FRIDAY
1. (CRAWL) LEADER DEVELOPMENT / VISUAL EXERCISE				Outline Staff Training; OPD on COA Analysis Process	Chess (Cognitive Exercise)
2. VISUAL EXERCISE / INTRODUCTION TO WG			Chess (Cognitive Exercise)	Chess (Cognitive Exercise)	(WG) Introduction
3. VISUALIZE / WIF TTX PREPARATION			Conduct WG	Conduct WG	Conduct WG; WIF TTX Assignments
4. (WALK) SEE OUR SELF TTX / LEADER DEVELOPMENT / TERRAIN TTX PREPARATION	Staff WIF TTX (Throughout the Week)				Leader Development Individual Staff Brief Their TTX; Terrain TTX Assignment
5. SEE THE TERRAIN TTX / LEADER DEVELOPMENT / ENEMY TTX PREPARATION	Staff Terrain TTX				Leader Development Individual Staff Brief Their Understanding of Terrain; Enemy TTX Preparation
6. SEE THE ENEMY TTX	S-2 (IPB) Develops Enemy Maneuver Template		S-2 Posts Enemy Template; SMEs Apply Enemy WIF Template	Leader Development Individual Staff Brief Their Enemy WIF	
7. (RUN) CONDUCT COA ANALYSIS	Deploy the Main CP	COA Analysis Number 1 (AAR)	COA Analysis Number 2 (AAR)	COA Analysis Number 3 (AAR)	AAR; Planning SOP Considerations and/or Updates

LEGEND:

(AAR) After Action Review
(COA) Course of Action
(CP) Command Post
(IPB) Intelligence Preparation of Battlespace
(OPD) Officers Professional Development
(S-2) Intelligence and Security
(SME) Subject Matter Expert
(SOP) Standard Operating Procedure
(TTX) Tabletop Exercise
(WIF) Warfighting Function
(WG) Wargame

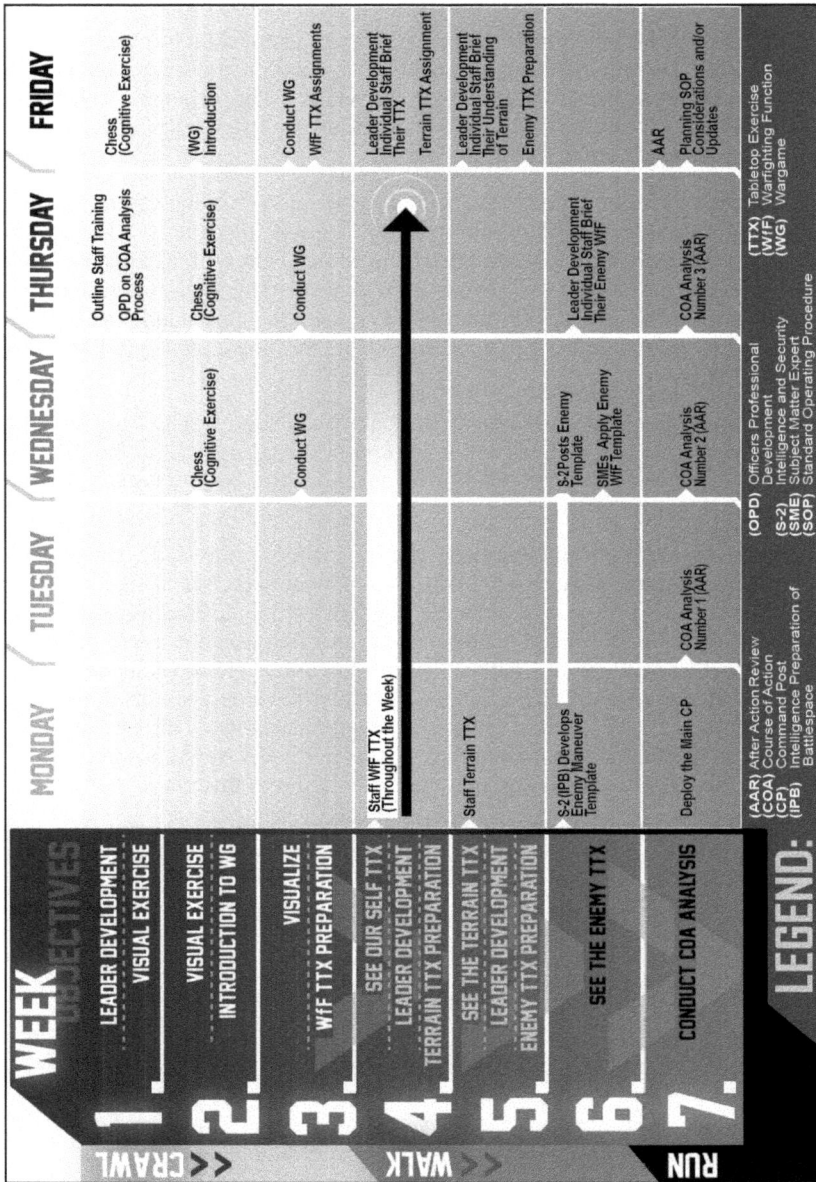

Figure 3-2. COA wargaming training plan (A way)

LIST OF SAMPLE COMMERCIAL WARGAMES

The following is a list of commercial wargames put together in coordination with the Directorate of Simulations Education, U.S. Command and General Staff College. This list does not constitute endorsement by the U.S. Army, and is not all-inclusive. There are many similar games out there, in addition to variations of the ones mentioned below. For commanders and staff that wish to explore these further, please contact the Directorate of Simulations Education, U.S. Command and General Staff College at (913) 684-3043 or (913) 684-3157.

Silver Bayonet by GMT Games

Description: Silver Bayonet is a turn-based board game depicting battalion and brigade battles of the 1st Cavalry Division against the North Vietnamese Army, with a heavy air assault component. It is not a counterinsurgency focused game. It incorporates morale and fatigue.

Training benefit: This game is useful for staff planning. It can be played as a campaign, and helps visualize brigade and battalion movement on a battlefield. It exercises the awareness of mental and physical readiness of troops, not just counting bayonet strength.

Complexity: Medium. A facilitator is required. Company-sized icons.

Players: 2-8

Time: 2-4 hours

Note: Effective for team play.

Silver Bayonet can be accessed at: https://boardgamegeek.com/ boardgame/7994/silver-bayonet-first-team-vietnam-1965

Alternate GMT games can be found at: https://www.gmtgames.com/

Combat Mission: Shock Force by Battlefront

Description: This is a series of real-time single player computer games replicating tactical actions at the company level. It has variants which include Stryker companies in Syria and Tank Teams in Ukraine. It is visually engaging, and includes individual vehicle and weapon icons.

Training benefit: This game is excellent for teaching planning and decision making at a tactical level with modern weapon systems. The game is relatively realistic and unforgiving of tactical blunders, and multiple scenarios allow it to be played multiple times.

Complexity: Medium. Requires some practice.

Players: 1

Time: Self-paced

Note: Have each player brief their planning efforts to their peers before playing. After the game, explore the various outcomes.

Combat Mission: Shock Force by Battlefront can be accessed at: http://www.battlefront.com/shock-force-2/

Alternate Battlefront games can be found at: http://www.battlefront.com/

Tactical Decision Games

Description: This is a "do-it-yourself" game. It provides a tactical (or other) dilemma, and gives a fixed amount of time to develop a solution. Players can work as individuals or in teams. Each team briefs their solution set and rationale, and is then critiqued by their peers. (Additional resources on the TDG website.)

Training benefit: This game creates confidence in developing situational awareness and decision making under pressure/time limits. The facilitated discussion/AAR can address almost any desired training or education focus areas.

Complexity: Easy

Players: 1 individual or up to 5 per team

Time: 4-6 hours. (This is dependent on the development by the facilitator, the executive officer.)

Note: There is a developer's workbook found in the supplemental resources, linked in Appendix C on page 87, which provides assistance in developing your own game. Develop and print a problem set and sketch. Incomplete or ambiguous information may be a deliberate part of the scenario. Provide specific guidance on what is to be briefed as part of the solution set (i.e., a COA sketch, assumptions, etc.). Set a time limit.

Note: If desired, the COA can be explored through informal COA analysis (wargaming) to examine it in more detail or explore branches, sequels, or longer-term consequences.

Tactical Decision Games can be accessed at: http://companyleader.themilitaryleader.com/tdg/

Pandemic by Z-Man Games

Description: This is a board game, which involves a cooperative team versus artificial intelligence. Players are trying to find cures for global

66

disease outbreaks while facilitating containment.

Training benefit: In addition to team building/ice breaking, the primary output of the game is that, during the AAR, the groups learn each other's character traits and how they think, while witnessing multiple concrete demonstrations of multiple mind traps, from group think to anchoring. This increases critical thinking, communication, and problem solving skills.

Complexity: Easy. It is a simple entry-level game for those uncomfortable with games.

Players: 2-4, plus an observer and/or facilitator

Time: Set up is 10 minutes. Training is 10 minutes. Play time is 1 hour.

Note: The maximum benefit of this game will derive from an introspective AAR discussing mind traps and group dynamics.

Pandemic by Z-Man Games cane be accessed at: https://www.zmangames.com/en/games/pandemic/

Alternate Z-Man games can be found at: https://www.zmangames.com/

Artemis Spaceship Bridge Simulator

Description: This game is a real-time computer simulation, requiring a cooperative team versus artificial intelligence. Players each have their own computers that simulate various battle stations on a Star Trek-like spaceship. Each position only has specific information, requiring information sharing to effectively understand a problem set and then develop and execute a solution.

Training benefit: This is a team-building exercise that focuses on information management to support decision making. Many of the problem sets present extremely ambiguous "black swan" events that truly challenge critical thinking and adaptability.

Complexity: Easy

Players: 6 (Additional games can be linked, allowing multi-ship team play or head-to-head adversarial play.)

Time: Practice is 30 minutes. Play time is 1-2 hours.

Note: Requires computers with access to a common server. Maximum training benefit accrues if a reflective AAR is conducted immediately afterwards.

Artemis Spaceship Bridge Simulator can be accessed at: https://artemisspaceshipbridge.com/

Flashpoint Campaigns: Red Storm Player's Edition by Matrix Games

Description: This is a turn-based, computer-driven, single-player game. This game replicates a tactical combined arms battalion minus to battalion plus-sized North Atlantic Treaty Organization (NATO) versus Warsaw Pact engagements in the mid-1980s. Icons typically represent platoons. The game has an easy, intuitive interface, incorporates fog of war, and is notable for including orders delay based on enemy electronic warfare jamming, among other things. It also incorporates troop fatigue and morale. Each scenario comes with a reasonable mission FRAGORD. It includes platoon and section-sized icons.

Training benefit: This game trains decision-making in ambiguous situations. It requires being able to envision yourself, the enemy, and terrain in relation to time and mission. It also teaches equipment capabilities, vulnerabilities, and hints at tactical best practices.

Complexity: Easy

Players: 1

Time: Set up is 5 minutes. Training/practice is 1 hour. Play-time is 1 hour.

Note: Most effective learning outcomes require some detailed planning prior to execution. A staff or individual can perform a reasonable MDMP on this. A time investment in conducting mission planning is well worth it. The game comes with multiple scenarios, many of which are non-U.S, but are still valid training scenarios and should be considered. Customizing scenarios is very time consuming.

Flashpoint Campaigns: Red Storm Player's Edition can be accessed at: https://www.matrixgames.com/products/471/details/flashpoint.campaigns:. redstorm

Alternate Matrix Games can be found at: https://www.matrixgames.com/

Main Battle Tank by GMT Games

Description: Main Battle Tank is a turn-based tactical board game. This game replicates tactical platoon to company-sized NATO versus Warsaw Pact engagements in the mid-1980s. Icons represent individual vehicles and squads. This game has optional rules allowing incorporation of more complex and realistic conditions. It has individual, vehicle, and squad icons.

Training benefit: This game trains decision making in ambiguous situations. It requires being able to envision yourself, the enemy, and terrain in relation to time and mission. It also teaches equipment capabilities, vulnerabilities, and tactical best practices.

Complexity: Medium

Players: 2-6 players

Time: Set up is 15 minutes. Training/practice is 45 minutes. Play-time is 1-2 hours. (Facilitator requires 2 hours to learn all rules.)

Note: A facilitator who understands all the rules, and can quickly answer questions and resolve combat outcomes, will greatly enhance the game. The facilitator can also incorporate realistic friction into the game. Time spent mission planning will greatly improve the learning benefit from the game. Fog of war can be incorporated by a facilitator. Double blind games, using multiple copies, take time and planning, but can greatly enhance realistic decision making. With some creativity, this can be adapted into a miniatures game.

Main Battle Tank can be accessed at: https://boardgamegeek.com/boardgame/157323/mbt-second-edition

Axis and Allies Global or 1942 by Milton Bradley

Description: This is a turn-based strategic board game. Teams of players play the major world powers of WW2 and compete militarily for resources, key terrain, and achieving key strategic conditions on a global scale. Players build naval, air, and ground forces with unique costs and capabilities.

Training benefit: This game is the perfect venue for reinforcing the principles of war. It is also a great way to train center of gravity analysis, and explores and teaches evaluating opportunity costs.

Complexity: Medium. Rules appear intimidating at first, but practice rounds can quickly gain players proficiency.

Players: 2-8

Time: 4-8 hours

Note: Practice rounds are necessary to understand the mechanics and develop feasible strategies.

Axis and Allies Global or 1942 can be accessed at: https://www.axisandallies.org/axis-and-allies-versions/

Kriegsspiel

Description: Kriegsspiel is the classic wargame that first brought wargaming into military training and education. It is a large board game played with blocks on enlarged military maps. It generally replicates Napoleonic meeting engagements from brigade to corps. It requires a facilitator. The unique aspect of this game is players are not allowed to communicate directly with each other or with their units. They must write notes to fellow players and provide written orders to their units, which they give to the facilitator. The facilitator then determines when to pass the message and how well the troops perform the actions. Krieggspiel uses battalion-sized icons.

Training benefit: This game is about effective communication, providing clear, succinct orders, and the exercise of command and control. Players have limited opportunities to influence their subordinates, so excellent guidance and intent is necessary.

Complexity: Easy

Players: 2-8 (Generally, however with 4-10 players, one or more facilitators are required. Given N players, you need $N+1$ umpires. One on the map, then one per player.)

Time: Set up is 20 minutes. Training/practice is 30 minutes. Play-time is 2 hours.

Note: Neither side's commander is allowed to see the game table. They are quarantined and must rely on the written reports from their subordinates for situational awareness. An AAR discussing command and control and communications post-game is where the most learning takes place. Ideally, with additional resources, this can be played as a double-blind game, which adds an altogether new aspect.

Kriegsspeil can be accessed at: https://boardgamegeek.com/boardgame/16957/kriegsspiel

Chess

Description: Chess is a classic tactical game that is about 1500 years old. It is a two-player game but, if played as a tournament, can generate good interaction and stimulates visualization.

Training benefit: This game is good for developing the conative skills of visualizing yourself and the enemy. Further, it forces the individual to see and analyze several moves in advance. Chess helps players to see the second and third ordered effects of their actions. With the pieces having different capabilities, it replicates combined arms, and stimulates the individual to gain a combined affect against his opponent. This is a reasoning and thinking game.

Complexity: Easy to difficult dependent on the experience of the individuals playing

Players: 2 (However, a chess tournament can involve the entire staff.)

Time: Set up is 5 minutes. Training/practice is 15 minutes. Play time depends on the 2 players, but on average 30-45 minutes.

Online chess can be accessed at: https://www.chess.com/

APPENDIX A

Technical Assistance Field Team Task Guide: Military Decisionmaking Process Collective Task Number 71-8-5111

Author's Note: Although Technical Assistance Field Team (TAFT) is no longer an active organization, the information found in this Task Guide is supported by Field Manual (FM) 6-0, *Commander and Staff Organization and Operations,* 05 MAY 2014 and the Digital Training Management System. The information was supported by Army Doctrine Reference Publication (ADRP) 5-0, *The Operations Process,* 17 MAY 2012, which has since been superceded by Army Doctrine Publication (ADP) 5-0, *The Operations Process,* as of 01 AUG 2019.

STEP 4 — COURSE OF ACTION ANALYSIS (WARGAMING)

Conditions: The staff is conducting or preparing to conduct operations. Communications are established with subordinates, adjacent units, and higher headquarters (HHQ). Command and Control (C2) Information Systems (INFOSYS) are operational and are passing information in accordance with tactical standard operating procedures (TSOPs). The command has received a warning order (WARNORD) from higher HQ, and is exercising mission command. Some iterations of this task should be performed in mission-oriented protective posture (MOPP) 4.

Standards: The staff analyzes a mission received from HHQ; develops, analyzes, and compares courses of action (COAs) against criteria of success and each other; selects the optimum COA to accomplish the mission; and produces and disseminates an operation plan/operation order (OPLAN/OPORD) to subordinates.

Note: Task steps and performance measures may not apply to every unit or echelon. Prior to evaluation, coordination should be made between evaluator and the evaluated units' HHQ to determine the task steps and performance measures that may be omitted.

Table A-1. Collective Tasks

Performance Step	Position	Performance Measures	Outputs
Coordinate COA analysis	XO	The Executive Officer (XO) coordinates actions of the staff during the wargame. He is the unbiased controller of the process, ensuring the staff stays on a timeline and achieves the goals of the wargaming session. In a time-constrained environment, the XO ensures that, at a minimum, the decisive operation is wargamed. (9-152)	• XO coordination and • COA development guidance.
Gather the tools	Staff Sections	The first task for *Course of Action (COA) Analysis* is to gather the necessary tools to conduct the wargame. The staff gathers the required tools to include: • Current running estimates, • Enemy templates and models, • Civil considerations and overlays, databases, and data files, • Modified combined obstacle overlay (MCOO) and terrain effects matrices, • Recording method, • Completed COAs, including graphics, • Method to post or display enemy and friendly unit symbols and other organizations, and • Map of the area of operations (AO). (9-123)	Tools and references for wargaming process. Step 1
List all friendly forces	Staff Officers	The staff lists the friendly forces with an emphasis on support relationships, constraints, and assets of all participants operating in the unit's AO. (9-124)	• Initial task organization and • Friendly forces list. Step 2
List assumptions	Staff Officers	The staff lists and reviews previous assumptions for continued validity and necessity. (9-125)	Revised list of assumptions. Step 3
List known critical events and decision points	Staff Officers	The staff lists known critical events and decision points (DP). Critical events include: • Those that trigger significant actions or decisions, such as commitment of an enemy reserve, • Those that are complicated actions requiring detailed study, such as a passage of lines, • Major events from the unit's current position through mission accomplishment, and • Possible reactions by civilians that potentially affect operations or that will require allocation of significant assets to account for essential stability tasks. The staff lists known critical events and decision points (DP). Critical events include DPs associated with: • The friendly force, • The status of ongoing operations, and • CCIR that describe what information the commander needs to make the anticipated decision. A decision point requires a decision by the commander. (9-126, 9-127)	Critical events with corresponding decision points. Step 4

Table A-1. Collective Tasks (Continued)

Performance Step	Position	Performance Measures	Outputs
Record the results of wargaming (*continued*)	Recorders	The synchronization matrix is a tool the staff uses to record the results of wargaming and helps them synchronize a course of action across time, space, and purpose in relationship to potential enemy and civil actions. (See table 9-3.) The sketch note method uses brief notes concerning critical locations or tasks and purposes. (See table 9-4.) These notes refer to specific locations or relate to general considerations covering broad areas. The commander and staff mark locations on the map and on a separate wargame work sheet. Staff members use sequential numbers to link the notes to the corresponding locations on the map or overlay. Staff members also identify actions by placing them in sequential action groups, giving each subtask a separate number. They use the wargame work sheet to identify all pertinent data for a critical event. They assign each event a number and title and use the columns on the work sheet to identify and list in sequence: • Units and assigned tasks, • Expected enemy actions and reactions, • Friendly counteractions and assets, • Total assets needed for the task, • Estimated time to accomplish the task, • The decision point tied to executing the task, • CCIR, • Control measures, and • Remarks. (9-137, 9-138, 9-172)	
Role play friendly commander	S-3 Officer or Commander	The S-3 role-plays the friendly maneuver commander. The S-3 is assisted by various staff officers such as the aviation officer and the engineer officer. (9-165)	S-3 / commander role plays friendly commander.
Execute friendly maneuver	S-3 Officer	The S-3 normally selects the technique for the wargame and role-plays the friendly maneuver commander. The S-3 executes friendly maneuver as outlined in the COA sketch and COA statement. (9-165)	Friendly maneuver conducted during the wargame.
Role play enemy commander	S-2 Officer	The S-2 role-plays the enemy commander. The S-2 develops critical enemy decision points in relation to the friendly COAs, projects enemy reactions to friendly actions, and projects enemy losses. When additional intelligence staff members are available, the S-2 assigns different responsibilities to individual staff members within the section for wargaming (such as the enemy commander, friendly S-2, and enemy recorder). By trying to win the wargame for the enemy, the S-2 ensures that the staff fully addresses friendly responses for each enemy COA. (9-164)	S-2 role plays enemy commander.
Develop critical enemy decision points	S-2 Officer	The S-2 develops critical enemy decision points in relation to the friendly COAs. (9-164)	Critical enemy decision points for each friendly COA.
Project and record enemy reactions to friendly action	S-2 Officer	The S-2 role-plays the enemy commander and captures the results of each enemy action and counteraction, as well as the corresponding friendly and enemy strengths and vulnerabilities. (9-164)	Enemy input to the wargame record.

75

Table A-1. Collective Tasks (Continued)

Performance Step	Position	Performance Measures	Outputs
Project enemy losses	S-2 Officer	The S-2 projects enemy losses. (9-164)	Projected enemy losses to the wargame record.
Identify IRs and recommend PIRs	S-2 Officer	The S-2 identifies information requirements (IR) and recommends priority intelligence requirements (PIR) that correspond to the decision points, including latest time information of value (LTIOV). These IRs are incorporated into the Information Collection Plan and graphics. (9-164)	IRs and PIR tied to decision points.
Refine event template and matrix for each COA	S-2 Officer	The S-2 refines the particular event template and matrix for each COA by adding named areas of interest (NAI) that support existing DPs, additional DPs, targeted areas of interest (TAI), and high-value targets (HVT). (9-164)	Refined event template and matrix for each COA.
Refine enemy SITTEMP for each COA	S-2 Officer	The S-2 refines each enemy situation template (SITTEMP) based on the wargaming results. (9-164)	Refined enemy SITTEMP for each COA.
Participate in targeting	S-2 Officer Targeting Team	The S-2, a key member of the Targeting Team, participates in targeting to select high-payoff targets (HPT) from HVTs identified during intelligence preparation of the battlefield (IPB). (9-164)	Participation in the Targeting Process.
Refine MCOO	S-2 Officer	The S-2 refines the modified combined obstacle overlay (MCOO) and terrain effects matrices. (9-164)	• MCOO refined. • Terrain effects matrices refined.
Refine weather products	S-2 Officer	The S-2 refines weather products that outline the critical weather impacts on operations. (9-164)	Weather products refined.
Assess adequacy of resources to provide human resources support	S-1 Officer	During the wargame, the S-1 assesses the personnel aspect of building and maintaining the combat power of units. This officer identifies potential shortfalls and recommends COAs to ensure units maintain adequate manning to accomplish their mission. As the primary staff officer assessing the human resources planning considerations to support sustainment operations, the S-1 provides human resources support for the operation. (9-168)	S-1 assessment of required human resources support.
Assess status of logistics functions	S-4 Officer	The S-4 assesses the status of all logistics functions required to support the COA, including potential support required to provide essential services to the civilians, and compares it to available assets. (9-169)	S-4 assessment of logistics status.
Assess logistics feasibility	S-4 Officer	The S-4 assesses the logistics feasibility of each wargamed COA. This officer determines critical requirements for each logistics function (classes I through VII, IX, and X) and identifies potential problems and deficiencies. (9-169)	S-4 assessment of logistics feasibility.
Identify potential shortfalls in support	S-4 Officer	The S-4 identifies potential shortfalls and recommends actions to eliminate or reduce their effects. While improvising can contribute to responsiveness, only accurately predicting requirements for each logistics function can ensure continuous sustainment. (9-169)	• Support shortfalls identified. • Recommendations for support.

Table A-1. Collective Tasks (Continued)

Performance Step	Position	Performance Measures	Outputs
Ensure movement support	S-4 Officer	The S-4 ensures that available movement times and assets support each COA. (9-169)	S-4 support to movement support.
Assess requirements, solutions, and concepts for each COA	S-5 Officer or S-3 Officer	The S-5 assesses warfighting requirements, solutions, and concepts for each COA; develops plans and orders; and determines potential branches and sequels arising from various wargamed COAs. The S-5 also coordinates and synchronizes warfighting functions (WFF) in all plans and orders. The planning staff ensures that the wargame of each COA covers every operational aspect of the mission. The members of the staff record each event's strengths and weaknesses and the rationale for each action. They complete the decision support template and matrix for each COA. They annotate the rationale for actions during the wargame and use it later with the commander's guidance to compare COAs. (9-165)	S-5 assessment of requirements, solutions, and concepts for each COA.
Assess COAs to determine required communications support	S-6 Officer	The S-6 assesses network operations, electromagnetic spectrum operations, network defense, and information protection feasibility of each wargamed COA. The S-6 determines communication systems requirements and compares them to available assets, identifies potential shortfalls, and recommends actions to eliminate or reduce their effects. (9-154)	S-6 assessment of communications support.
Assess informational aspects of each COA	IO Officer or S-3 Officer	The Information Operations (IO) Officer (or S-3 if there is no IO Officer) assesses how effectively the operations reflect the information-related capabilities (IRC), the effectiveness of capabilities to execute (deliver) IRC in support of each wargamed COA, and how IRC impact various audiences of interest and populations in and outside the AO. The IO Officer also integrates IO with cyber-electromagnetic activities (CEMA) with themes and messages. (9-155)	IO Officer assessment of informational aspects of each COA.
Assess civil considerations of each COA	S-9 Officer or S-3 Officer	The S-9 ensures each wargamed COA effectively integrates civil considerations (the "C" of METT-TC). The civil affairs operations officer considers not only tactical issues but also sustainment issues. This officer assesses how operations affect civilians and estimates the requirements for essential stability tasks commanders might have to undertake based on the ability of the unified action. Host-nation support and care of dislocated civilians are of particular concern. The civil affairs operations officer's analysis considers how operations affect public order and safety, the potential for disaster relief requirements, noncombatant evacuation operations, emergency services, and the protection of culturally significant sites. This officer provides feedback on how the culture in the AO affects each COA. If the unit lacks an assigned civil affairs operations officer, the commander assigns these responsibilities to another staff member. (9-156)	Civilian considerations integrated into COAs.

Table A-1. Collective Tasks (Continued)

Performance Step	Position	Performance Measures	Outputs
Assess fire support aspects of each COA	FSO Targeting Team	The FSO (Chief of Fires) assesses the fire support feasibility of each COA. The FSO develops a proposed high-payoff target list (HPTL), target selection standards (TSS), and an attack guidance matrix (ATG). The FSO works with the S-2 to identify named and targeted areas of interest, HPTs, and additional events that may influence the positioning of field artillery and air defense artillery (ADA) assets. The FSO should offer a list of possible defended assets for ADA forces and assist the commander in making a final determination about asset priority. (9-166)	• HPTL, • TSS, • AGM, and • ADA defended assets list.
Assess protection element requirements	Protection or S-3	The chief of protection assesses protection element requirements, refines essential elements of friendly information (EEFI), and develops a scheme of protection for each wargamed COA. The chief: • Refines the critical asset and the defended asset lists, • Assesses hazards, • Develops risk control measures and mitigation measures of threats and hazards, • Establishes personnel recovery coordination measures, • Implements operational area security to include security of lines of communications, antiterrorism measures, and law enforcement operations, • Ensures survivability measures reduce vulnerabilities, and • Refines chemical, biological, radiological, and nuclear operations. (9-167)	Assessment of protection requirements.
Perform risk assessment	Staff Officers	The staff continually assesses the risk to friendly forces balancing between mass and dispersion. When assessing the risk of weapons of mass destruction to friendly forces, planners view the target that the force presents through the eyes of an enemy target analyst. They consider ways to reduce vulnerability and determine the appropriate level of mission-oriented protective posture consistent with mission accomplishment. They identify hazards, assess their risk, develop controls for them, and determine residual risk, all in accordance with the Risk Management (RM) Process detailed in FM 5-19. (9-145; FM 5-19)	Assessment of risk for each COA.

Conduct a wargame briefing (optional) (*continued on the next page*)	Commander XO Staff Officers	Time permitting, the staff conducts a wargame brief to the commander (optional):
		• HHQ mission, commander's intent, and Military Deception (MILDEP) Plan,
		• Updated IPB,
		• Friendly and enemy COAs that were wargamed:
		  ➢ critical events,
		  ➢ possible enemy actions and reactions,
			Internal staff wargaming briefing and results of wargame.
			
			Step 8

Table A-1. Collective Tasks (Continued)

Performance Step	Position	Performance Measures	Outputs
Conduct a wargame briefing (optional) (*continued*)	Commander XO Staff Officers	Time permitting, the staff conducts a wargame brief to the commander (optional): • Friendly and enemy COAs that were wargamed: ➢ possible media impacts, ➢ possible impacts on civilians, ➢ modifications to the COAs, ➢ strengths and weaknesses, and ➢ results of the wargame. • Assumptions, and • Wargaming technique used. (9-149)	**Step 8**
Participate in the wargame briefing (optional)	Commander XO Staff Officers	The staff attends the optional wargame briefing to the XO or commander and participates, as required. (9-149)	Participation in wargame briefing (optional).
Refine CCIR	Commander Staff Officers	The list of commander's critical intelligence requirements (CCIR) constantly changes. The commander constantly modifies and refines CCIR and incorporates them into the IC Plan. (1-24, ADRP 5-0; Table 9-5)	Refined CCIR.
Revise and update running estimates	Staff Sections	The staff assesses the currency of running estimates and determines those needing updates. The task of developing and updating running estimates continues throughout the Military Decisionmaking Process (MDMP) and the Operations Process. (9-19)	Revised running estimates.
Direct, supervise, and coordinate staff planning	XO	The XO is a key participant in the MDMP. The XO manages and coordinates the staff's work and provides quality control during the MDMP. To effectively supervise the entire process, this officer has to clearly understand the commander's intent and guidance. The XO provides timelines to the staff, establishes briefing times and locations, and provides any instructions necessary to complete the plan. (2-32, 9-10)	XO direction, supervision, and coordination.

Table A-2. Supporting Individual Tasks. Participate in COA Analysis and Wargaming

Performance Step	Performers	Outputs
Gather required tools	Staff Officer	The staff officer participates in *Course of Action (COA) Analysis (Wargaming)* to identify the advantages and disadvantages of each COA. The staff officer gathers the required tools: • Running estimates, • Threat templates and models, • Civil considerations and overlays, databases, and data files, • Modified combined obstacle overlay (MCOO) and terrain effects matrices, • Recording method, • Completed COAs, including graphics, • Method to post or display enemy and friendly unit symbols and other organizations, and • Map of the area of operations (AO).
List friendly forces	Staff Officer	The staff officer participates in *Course of Action (COA) Analysis (Wargaming)* to identify the advantages and disadvantages of each COA. The staff officer lists friendly forces with an emphasis on support relationships, constraints, and assets of all participants operating in the unit's AO.
List and review assumptions	Staff Officer	The staff officer participates in *Course of Action (COA) Analysis (Wargaming)* to identify the advantages and disadvantages of each COA. The staff officer lists and reviews with the commander previous assumptions for continued validity and necessity.
List known critical events and decision points	Staff Officer	The staff officer participates in *Course of Action (COA) Analysis (Wargaming)* to identify the advantages and disadvantages of each COA. The staff officer lists known critical events and decision points (DPs), to include: • Critical events that: ➢ trigger significant actions or decisions, such as commitment of an enemy reserve, ➢ are complicated actions requiring detailed study, such as a passage of lines, ➢ includes major events from the unit's current position through mission accomplishment, and ➢ includes possible reactions by civilians that potentially affect operations or that will require allocation of significant assets to account for essential stability tasks. • Decision points (DP) associated with: ➢ friendly force, ➢ status of ongoing operations, and ➢ CCIR that describe what information the commander needs to make the anticipated decision.
Select wargaming method (*continued on the next page*)	Staff Officer	The staff officer selects the wargaming method from one of the following methods or by developing a different technique: • Belt method – dividing the AO into belts (areas) running the width of AO, • Avenue-in-depth method – focusing on one avenue of approach at a time, beginning with the decisive operation or main effort, and/or • Box method – detailed analysis of critical areas such as an engagement area (EA), a wet gap crossing site, or a landing zone.

Table A-2. Supporting Individual Tasks. Participate in COA Analysis and Wargaming (Continued)

Performance Step	Performers	Outputs
Select wargaming method (*continued*)	Staff Officer	The staff officer selects the wargaming method from one of the following methods or by developing a different technique: During stability operations: • The belt method divides the COA by events, objectives, or events and objectives in a selected slice across all lines of effort (LOE), • The avenue-in-depth method can be modified by focusing on wargaming a LOE by reviewing the relationship among events or objectives on all LOEs with respect to events in the selected line, and • The box method focuses on a specific objective along a LOE.
Select technique to record and display results	Staff Officer	The staff officer selects a technique to record and display wargaming results: • Synchronization matrix and • Sketch note technique.
Execute action / reaction / counteraction analysis	Staff Officer	The staff officer participates in wargaming the COAs and assesses the results. He/she executes action / reaction / counteraction analysis through each COA's selected events.
Assess all possible forces	Staff Officer	The staff officer participates in the COA wargaming and assesses the results. He/she considers all possible forces, including actions of civilians in the AO and templated enemy outside the AO that can influence the operation. Verifies each COA results in a completed synchronization matrix. Lists assets used in appropriate columns of worksheets and lists totals in the assets column, while not considering any assets lower than two command echelons below the staff.
Evaluate friendly moves	Staff Officer	The staff officer participates in the COA wargaming and assesses the results. He/she evaluates each friendly move to determine assets and actions required to defeat the threat at that point, or to accomplish stability tasks.
Considers branches to the plan	Staff Officer	The staff officer participates in the COA wargaming and assesses the results. He/she considers branches to the plan that promote success against likely enemy counteractions or unexpected civilian reactions.
Identify items that require further analysis	Staff Officer	The staff officer participates in the COA wargaming and assesses the results. He/she identifies situations, opportunities, or additional critical events that require further analysis by the staff. He/she examines the following areas in detail, to include, but not limited to: • All friendly capabilities, • All enemy capabilities and critical civil considerations that impact operations, • Global media responses to proposed actions, • Movement considerations, • Closure rates, • Lengths of columns, • Formation depths, • Ranges and capabilities of weapons systems, and • Desired effects of fires.
Assess risk to friendly forces	Staff Officer	The staff officer participates in the COA wargaming and assesses the results. He/she assesses risk to friendly forces and develops ways to mitigate those risks.
Identify WFFs required to support the concept of operations	Staff Officer	The staff officer participates in the COA wargaming and assesses the results. He/she identifies warfighting function (WFF) assets required to support the concept of the operations.

Table A-2. Supporting Individual Tasks. Participate in COA Analysis and Wargaming (Continued)

Performance Step	Performers	Outputs
Recommend priorities	Staff Officer	The staff officer participates in the COA wargaming and assesses the results. He/she recommends priorities if requirements exceed available assets, based on the situation, commander's intent, and planning guidance.
Incorporate results of analysis into recording method	Staff Officer	The staff officer participates in the COA wargaming and assesses the results. He/she performs any additional analysis and incorporates the results into the wargame recording method.
Consider coverage by the global media	Staff Officer	The staff officer considers the diverse kinds of coverage of unfolding events and their consequences in the global media.
Refine or modify areas as a result of the wargame	Staff Officer	The staff officer participates in the COA wargaming and assesses the results. He/she refines or modifies the following: • Each COA, to include identifying branches and sequels that become on-order or be-prepared missions, • Locations and times of DPs, • Enemy event template and matrix, • Task organization, including forces retained in general support, • Control requirements, including control measures and updated operational graphics, and • CCIR, EEFI, and other IRs, including LTIOV, and incorporate them into the Information Collection Plan.
Identify areas as a result of the wargame	Staff Officer	The staff officer participates in the COA wargaming and assesses the results. He/she identifies the following: • Key or decisive terrain and determining how to use it, • Tasks the unit retains and tasks assigned to subordinates, • Likely times and areas for enemy use of weapons of mass destruction (WMD) and friendly chemical, biological, radiological, and nuclear (CBRN) defense requirements, • Potential times or locations for committing the reserve, • The most dangerous enemy COA, • The most likely enemy COA, • The most dangerous civilian reactions, • Locations for the commander and command posts (CP), • Critical events, • Requirements for support of each WFF, • Effects of friendly and enemy actions on civilians, infrastructure, and on military operations, • Confirmed locations of NAIs, TAIs, DPs, and IRs needed to support them, • The strengths and weaknesses of each COA, • Hazards, assessing their risk, developing controls of them, and determining residual risk, and • The coordination required for integrating and synchronizing interagency, host nation (HN), and nongovernmental involvement.
Analyze areas as a result of the wargame	Staff Officer	The staff officer participates in the COA wargaming and assesses the results. He/she analyzes the following: • Potential civilian reactions to operations, • Potential media reaction to operations, and • Potential impacts on civil security, civil control, and essential services in the AO.

Table A-2. Supporting Individual Tasks. Participate in COA Analysis and Wargaming (Continued)

Performance Step	Performers	Outputs
Develop areas as a result of the wargame	Staff Officer	The staff officer participates in the COA wargaming and assesses the results. He/she develops the following: • DPs, • Synchronization matrix, • Decision support template and matrix, • Solutions to achieving minimum essential stability tasks in the AO, • Information Collection Plan and graphics, • Initial information-related capabilities (IRC) and cyber electromagnetic activities (CEMA), and • Fires, protection, and sustainment plans and graphic control measures.
Determine areas as a result of the wargame	Staff Officer	The staff officer participates in the COA wargaming and assesses the results. He/she determines the following: • Requirements for military deception (MILDEP) and surprise, • Timing for concentrating forces and starting the attack or counterattack, • Movement times and tables for critical assets, including mission command systems (MCS), • Estimated duration of the entire operation and each critical event, • Projected percentage of enemy forces defeated in each critical event and overall, • Percentage of minimum essential tasks that the unit can or must accomplish, • Media coverage and impact on key audiences, • Targeting requirements in the operation, to include identifying or confirming HPTs and establishing attack guidance, and • Allocation of assets to subordinate commanders to accomplish their missions.

Participate in the COA briefing	Staff Officer	The staff officer participates in the conduct of a wargame brief to the commander (OPTIONAL), time permitting, to include, but not limited to:
		• HHQs mission, commander's intent, and MILDEP Plan,
		• Updated IPB,
		• Assumptions,
		• Friendly and enemy COAs that were wargamed, to include:
		  ➤critical events,
		  ➤possible threat actions and reactions,
		  ➤possible impacts on civilians,
		  ➤possible media impacts,
		  ➤modifications to the COA,
		  ➤strengths and weaknesses, and
		  ➤results of the wargame.
		• Wargaming technique used.

APPENDIX B

References

Adkinson, F.L. Personal Interview on Observations on COA Analysis from the Joint Maneuver Readiness Center. December 10, 2018.

Army Doctrine Publication (ADP) 5-0. *The Operations Process.* 31 JUL 2019.

ADP 6-0. *Mission Command: Command and Control of Army Forces.* 31 JUL 2019.

Benet, S.V. (1961). *John Brown's Body.* (pp. 27-28). New York, NY: Dramatists Play Service, Inc.

Burggrabe, Reec. Personal Interview on Doctrinal Updates and TTPs for COA Analysis. February 20, 2019.

Field Manual (FM) 6-0. *Commander and Staff Organization and Operations.* 05 MAY 2014. C1 incorporated 22 APR 2016.

Frame, J. E. (1997). *Gazing into the Crystal Ball Together: Wargaming and Visualization for the Commander and Staff.* Fort Leavenworth, Kansas: School of Advanced Military Studies, U.S. Army Command and General Staff College.

Frank, Anders (2012). "Gaming the Game: A Study of the Gamer Mode in Educational Wargaming." *Simulation and Gaming* 43, no. 1: 118-132.

Garra, N. A. (2004). *Wargaming a Systematic Approach, Discover the Fast Effective way to Supercharge your Battle Staff.* Sierra Vista, Arizona: S2 Company.

Infantry School Staff. (1939) Terrain. *Infantry in Battle* (pp. 69-78). Washington D.C : The Infantry Journal Incorporated.

Kretchik, W. E. (1991). *The Manual Wargaming Process: Does our Current Methodology Give Us the Optimum Solutions? A Monograph for the School of Advanced Military Studies.* Fort Leavenworth, Kansas: School of Advanced Military Studies.

Krueger, R. L. (2018). *CTCD Quarterly Bulletin 01-09 Wargame Planning Considerations.* Fort Polk, LA: Joint Readiness Training Center.

Leader Training Program Coach, N. T. Personal Interview on Observations and Trends on Unit's Conduct of Course of Action Analysis (War-gaming). March 3, 2019.

Longacre, E.G. (2000). *Lincoln's Calvarymen: A History of the Mounted Forces of the Army of the Potomac.* (p. 183) Mechanicsburg, PA: Stackpole Books.

McConnell, R.A. (2020). "Connecting the Dots: Developing Leaders who can Turn Threats into Opportunities." *Military Review*

McConnell, R. A., and Gerges, M. T. (2018). *Seeing the Elephant, Improving Leader Visualization Skills through Simple War Games.* Military Review Online Exclusive.

McConnell, R. A., Gerges, M., Dalbey, J., Dial, T., Hodge, G., Leners, M.,... Schoof, P. (2018). *The Effect of Simple Role-Playing Games on the Wargaming Step of the Military Decisionmaking Process (MDMP) A Mixed Method Approach.* Fort Leavenworth, Kansas: U.S. Army Command and General Staff College.

Ministry of Defence. (2017). *The Wargaming Handbook.* The Development, Concepts and Doctrine Centre: Ministry of Defence Shrivenham.

Napoleon 1. (1910). *The Corsican: A Diary of Napoleon's Life in His Own Words.* Boston, MA: Houghton Mifflin.

Pray, J. L.-O. Personal Interview on Observations on COA Analysis from the Joint Maneuver Readiness Center. December 10, 2018.

Reinwald, B. R. (2000). "Tactical Intuition." *Military Review* 80, no. 5. (pp. 80-88).

Rubel, Robert C. (2006). "The Epistemology of War Gaming." *Naval War College Review* 59, no. 2, Article 8.

Sallot, Steven. Personal Interview on Applying Commercial Games to Military Education. January 15, 2019.

LTC Smith, C. R. (1863). Brigadier General John Buford, U.S. Army, commanding First Division: Battle of Gettysburg. In United States War Department (Ed.), *The War of the Rebellion: A Compilation of the Official Records of the Union and Confederate Armies.* (Series 1, V. 27–Part 1) (pp. 927-993). Washington, Government Printing Office.

Scott, H. D. (1992). *Time Management and the Military Decision Making Process. A Monograph for the School of Advances Military Studies, U.S Command and General Staff College.* Fort Leavenworth, Kansas: School of Advanced Military Studies.

LTC Spurlin, D. and LTC Green M. (2017). Demystifying the Correlation of Forces Calculator. *Infantry Magazine January-March 2017* (pp. 14-17).

Stackpole, E. J. (1982). *They Met At Gettysburg*. Harrisburg, PA: Stackpole Books.

Stallings, P. A. (1992). *What To Do, What To Do? Determining a Course of Action at the Operational Level of War. A Monograph for the School of Advanced Military Studies*. Fort Leavenworth, KS: School of Advanced Military Studies

Stanley, E. B. (2000). *Wargames, Training, and Decision-Making, Increasing the Experience of Army Leaders*. Fort Leavenworth, KS: School of Advances Military Studies, U.S. Army Command and General staff College.

Sun Tzu, T. B. (1976). *SUN TZU, The Art of War*. New York: Printed in the United States of America.

Training Circular (TC) 6-0. *Training the Mission Command Warfighting Function*. 21 DEC 2017.

Training Analysis Feedback Team. (2014). Military Decision Making Process (MDMP) COA Analysis (War-gaming). *Task Guide MDMP*, (pp. 80-100). Fort Leavenworth, Kansas: The Flemming Group, LLC.

U.S. Army Command and General Staff College. (1991). Student Test 100-9. *Techniques and Procedures for Tactical Decision-making*. Fort Leavenworth, Kansas: U.S. Army Command and General Staff College.

U.S. Army Command and General Staff College. (1993). Student Text 100-9. *The Tactical Decision Making Process*. Fort Leavenworth, Kansas: U.S. Army Command and General Staff College.

U.S. Army Command and General Staff College. (1989). Student Text 100-9. *The Command Estimate*. Fort Leavenworth, Kansas: U.S. Army Command and General Staff College.

U.S. Army Command and General Staff College. (1992). Student Text 100-9. *The Command Estimate Process*. Fort Leavenworth, Kansas: U.S. Command and General Staff College.

Wood, W. (1995). *Leaders and Battles, The Art of Military Leadership*. Novato, CA: Presidio Press.

Wood, W. J. (1990). *Battles of the Revolutionary War 1775-1781*. Chapel Hill, NC: Algonquin Books of Chapel Hill.

Supplemental Resources

These supplemental resources provide additional information to enhance this handbook and assist in educating and training the staff.

These resources can be accessed on the Center for Army Lessons Learned restricted website at: https://call2.army.mil/toc.aspx?document=17879 (Common Access Card (CAC) login required)

The resources found at this link are organized into three files. The first file is a course of action (COA) analysis video from the School of Advanced Military Studies (SAMS), which contains a three-part video of a SAMS staff group conducting COA analysis (wargaming). The second file contains professional reading — ten articles and a leader development program wargaming handbook. The third file contains further useful documents for staff training.

FILE ONE: COURSE OF ACTION ANALYSIS (WARGAMING) VIDEOS

Dr. Bruce Stanley leads parts one and two of this COA analysis video series. Part one is a COA analysis walk through, lasting an hour and twenty-three minutes. It focuses on setting up and friendly actions for wargaming. Part two is an hour and ten minutes, and focuses on the first two tactical events: the forward passage of lines/security zones, enemy reactions, friendly counteractions, and recap. The third video is an hour and thirty-two minute class focused on COA analysis (wargaming) — large-scale combat operations academics. The following two links from the National Training Center and the Joint Readiness Training Center provide additional video examples of wargaming. (CAC Access Required.)

National Training Center: https://www.milsuite.mil/book/leadercasts/7582

Joint Readiness Training Center: https://atn.army.mil/joint-readiness-training-center-(jrtc)/joint-readiness-training-center-(jrtc)

FILE TWO: PROFESSIONAL READINGS

Correlation of Forces. The Quest for a Standardized Model
Major David R. Hogg

This study is an examination of how to measure combat power. The different methods to measure combat power range from a numerical count (bean count) to the subjective and objective analysis of individual weapon systems and/or units. The critical base to any correlation of forces model is the values associated with the weapon systems or units. Four different correlation of forces models are examined using specific criteria. The models studied are: The National Training Center model, the Command and General Staff College model, the Theater Analysis model, and the Historical Evaluation and Research Organization model. The criteria applied to each of these models are: flexibility, simplicity, definable values, and the ability to provide at least a 90 percent solution. The conclusion of this study is that a standardized model is needed, and that the model should be based on individual weapon system values (using operational lethality index factors).

Development and Assessment of Battlefield Visualization Training for Battalion Commanders
Scott B. Sahdrick, David Manning, James Bell, Dennis K. Leedom, and Carl W. Lickteig

This article focuses on visualization. The art and science of developing situational understanding, determining a desired end state, and envisioning how to move the force from its current state to the desired end state is critical to successful battle command (mission command). Today's Army does not have the most effective method for developing expert visualization skills. Recent research on expertise indicates that experience alone, be it real or in simulated battle, is not adequate. Instead, expertise is more likely to be attained through a combination of education, training, practice, and experience.

From the Interservice/Industry Training, Simulation, and Education Conference (J1TSEC 2008).

Seeing the Elephant, Improving Leader Visualization Skills through Simple War Games
LTC Richard A. McConnell, DM, and LTC Mark T. Gerges, PhD

This is a summary of an exhaustively detailed academic paper titled *The Effect of Simple Role-Playing Games on the War-gaming Step of the Military Decisionmaking Process (MDMP): A Mixed Methods Approach*, previously published in *Developments in Business Simulation and Experimental Learning: Proceedings of the Annual Association for Business Simulation and Experiential Learning Conference* 45 (2018). For those interested in seeing the entire paper, it can be accessed at: https://journals.tdl.org/absel/index.php/absel/article/view/3200/3127

Connecting the Dots: Developing Leaders who can Turn Threats into Opportunities
R.A. McConnell

Connecting the Dots is the sequel to *Seeing the Elephant*. It is the latest milestone in a multiple year and publication journey, attempting to understand visualization, its components, and how to improve this vital skill. It was written to present literature and analysis that could serve as a foundation for further research and publication, which could result in recommendations to improve leader visualization skills through deliberate practice.

Tactical Intuition
Major Brian R. Reinwald

This article focuses on visualizing and improving intuition.

The Manual Wargaming Process: Does our Current Methodology Give us the Optimum Solution
Major Walter E. Kretchik

This monograph analyzes the manual wargaming portion of the U.S. Army's decision-making cycle, in order to determine if the process deduces the optimal COA.

Time Management and the Military Decisionmaking Process
Harry D. Scott, Jr.

This monograph analyzes the military decisionmaking process in terms of time management, in order to determine if a timeline will expedite the process. It begins by establishing the importance of time and time management in planning. There is also a general discussion of time, an explanation of the Army's one-third to two-thirds rule, and a synopsis of the deficiencies and recommendations for improvement of units' execution of the military decisionmaking process during rotations to the National Training Centers. The monograph concludes with the advantages and disadvantages of utilizing a timeline. The end result is that the advantages of a timeline far outweigh the disadvantages.

Wargame Planning Considerations
LTC Roy Krueger

This is a joint readiness training center, CTC quarterly bulletin, focused on providing ways to improve COA analysis.

Wargames, Training, and Decision-Making: Increasing the Experience of Leaders
Major Bruce E. Stanley

This monograph examines the question, can commercial computer wargames increase the experience level and decision-making abilities of Army leaders? Additionally, the monograph looks at three secondary questions. How can the Army use computer wargames to increase experience and decision-making? Why should the Army use computer wargames? And, what are the benefits of computer wargames for the Army? This monograph shows that computer wargames, when used consistently by Army leaders, can increase experience and decision-making skills. Like any training, computer wargames must be used repetitively to achieve results.

Gazing Into the Crystal Ball Together: Wargaming and Visualization for the Commander and Staff
Major John R. Frame

This monograph discusses the importance of the commander and staff wargaming together. Wargaming is a critical visualization event where the participants develop detailed images of the operation. Wargaming allows the commander and staff to build a common vision and understanding of battle.

Demystifying the Correlation of Forces Calculator
LTC Dale Spurlin and LTC Matthew Green

This article describes the development of the correlations of forces and means calculator currently in use with the Department of Army Tactics at the U.S. Army Command and General Staff College. It addresses the methodology used to determine the values, suggests appropriate uses of the tool, and suggests some ideas for adding professional judgment to the results.

The Epistemology of Wargaming
Robert C. Rubel

This article discusses potential problems that may occur if individuals or groups do not adhere to the principles and realities of wargaming. Wargames can be useful for planning and decision-making, but they can also produce "valid looking garbage." There are principles that can help users and analysts tell the difference and avoid the pitfalls, but if game results are to earn the confidence they are now given, the craft of gaming must become a profession.

Gaming the Game: A Study of the Gamer Mode in Educational Wargaming
Anders Frank

This article discusses problems that can occur if players do not focus on their actual training goals while using off-the-shelf wargames. One risk associated with the use of games in training and education is that players start to "game the game," instead of focusing on their learning goals. The term "gamer mode" is proposed to describe this attitude. A player with a gamer mode attitude strives to achieve goals that are optimal for winning the game, but suboptimal with respect to the educational objectives. In this study of cadets playing an educational wargame to learn ground warfare tactics, the author examined the occurrence of gamer mode. The results showed that gamer mode emerged sporadically throughout all analyzed sessions. Cadets' understanding of the wargame was different from what the instructors expected. This study discusses why it is important to avoid situations where gamer mode can emerge, and also speculates on the sources that generate this attitude — the game itself, the educational setting, and the participants' previous experiences.

FILE THREE: SUPPLEMENTAL TRAINING MATERIAL

Design and Delivery of Tactical Decision Games Sand Table Exercises: A Tool Box Reference

This workbook is to assist leaders in the design and delivery of tactical decision games and sand table exercises. The first part of this workbook focuses on the design of specific exercises, while the second part focuses on delivery techniques that will enhance the success and effectiveness of the exercises. Tactical decision games and sand table exercises, when properly designed and delivered, will allow the staff to practice situational assessment, to consider, select, and war-game COAs, and to practice communicating those decisions.

Based on *The How To of Tactical Decision Games* by Major John F. Schmitt, United States Marine Corps, 1994. Marine Corps University Publication.

Task Guide and Military Decisionmaking Process Reference

This task guide provides detailed tasks, conditions, standards, performance measures, inputs, and outputs for the entire military decisionmaking process. This is helpful for planning and assessing.

Product of the Technical Assistance Field Team (TAFT), Fort Leavenworth, KS.

Integrated Staff Planning Matrix: Receive the Mission Through Wargaming

This is a helpful graphic training aid on wargaming.

LTC Richard B. Averna (U.S. Army, Retired)

Correlation of Forces Calculator: Automated and Analog, Version 2017.01

This can be helpful in determining relative force ratios. Once this Excel document is opened, directions for usage can be found under the second tab.

Provided by the Tactics Division at the Command and General Staff College (CGSC), in addition to Dr. James E. Sterrett and Mr. Michael B. Dunn, Directorate of Simulation Education, U.S. Army, Command and General Staff College.

Wargaming Big Picture
Jacob A. Mong

This is a PowerPoint file that provides a quick reference for Step 4 of COA Analysis (Wargaming), and can be used to generate discussion. It provides the following discussion points: input/process/output; example of room setup; a wargame flow chart; decision-making in execution; execution decisions; and adjustment decisions.

Unit Icon Template

The unit icon template for the correlation of forces calculator is a grouping of enemy and friendly graphic unit icons that can be developed to reflect the table of organization and equipment for use during the wargame enemy.

Developed by CGSC Students, Fort Leavenworth, KS.

Sub-Folder

This sub-folder includes examples of synchronization and decisions matrices that can be adapted for use.

These examples are taken from the National Training Center and the Joint Readiness Training Center.

Milton Keynes UK
Ingram Content Group UK Ltd.
UKHW020639041223
433752UK00017B/868